RANDOM SEQUENTIAL PACKING OF CUBES

RANDOM SEQUENTIAL PACKING OF CUBES

Mathieu Dutour Sikirić
Ruđer Bošković Institute, Croatia

Yoshiaki Itoh
The Graduate University for Advanced Studies, Japan
& The Institute of Statistical Mathematics, Japan

NEW JERSEY · LONDON · SINGAPORE · BEIJING · SHANGHAI · HONG KONG · TAIPEI · CHENNAI

Published by

World Scientific Publishing Co. Pte. Ltd.

5 Toh Tuck Link, Singapore 596224

USA office: 27 Warren Street, Suite 401-402, Hackensack, NJ 07601

UK office: 57 Shelton Street, Covent Garden, London WC2H 9HE

Library of Congress Cataloging-in-Publication Data
Dutour Sikirić, Mathieu.
    Random sequential packing of cubes / by Mathieu Dutour Sikirić & Yoshiaki Itoh.
    p. cm.
    Includes bibliographical references.
    ISBN-13: 978-981-4307-83-3 (hardcover : alk. paper)
    ISBN-10: 981-4307-83-1 (hardcover : alk. paper)
    1. Combinatorial packing and covering. 2. Sphere packings. I. Itoh, Yoshiaki, 1943–
II. Title.
    QA166.7.D88 2011
    511'.6--dc22

                                                              2010027617

British Library Cataloguing-in-Publication Data
A catalogue record for this book is available from the British Library.

Printed in Singapore.

To Maja and Yuri

Preface

A cube packing is a family of translates of the unit cube $[0,1]^n$ such that the intersection of any two of them has empty interior. If one starts from the empty set and add translates, chosen at random, of the unit cube sequentially until there is no space to pack any more, then one gets a non-extensible cube packing.

The Flory model is perhaps the first, such sequential random packing. It was introduced in Polymer Chemistry to model the behavior of catalyst and it considers integral translates of an interval $[0,2]$ inside another interval $[0,n]$ and asks for the limit packing density which represents then the efficiency of the catalyst. Another classical such model is the random parking problem, where drivers park their cars of length 1 at random in a parking place interval $[0,x]$. Such models have wide applications to statistical physics.

The first major result in the field is the computation of the limit density C_R of the random parking problem by [Rényi (1958)]. [Palásti (1960)] conjectured that the average density for the sequential random packing of $[0,1]^d$ in $[0,x]^d$ converges to C_R^d when x goes to ∞. The existence of the limiting packing density was shown only recently by [Penrose (2001)] but the rigorous computation of the limit packing density is not known for dimension greater than 1. To estimate this d-dimensional limit packing density, a classic method is the sampling procedure or Monte-Carlo simulation of the random sequential packing itself. The results of those simulations do not support Palásti's conjecture on the value of the limit packing density [Blaisdell and Solomon (1970)] and currently the limiting packing density is not known rigorously in dimension $d > 1$.

The 1-dimensional random sequential packing problem can be solved within the framework of classical mathematical analysis. For example in order to study the recursion formula to get the limiting packing density by [Rényi (1958)], we introduce delay integral equations, for which we apply Laplace transform and then finally a Tauberian theorem which we prove completely. The minimum of gaps gives an interesting nonlinear delay integral equation [Itoh (1980)] whose analysis is amenable by tools of classical analysis which we introduce appropriately when needed. The alternative methodology of [Dvoretzky and Robbins (1964)] is also introduced as well as the related central limit theorem. We consider then a unified Kakutani-Rényi model, which interpolates between the Kakutani interval splitting and the Rényi sequential random packing [Komaki and Itoh (1992)]. We also consider a spin variant of the packing problem, where cars can park with spin-up or spin-down and compute the corresponding limit packing density [Itoh and Shepp (1999)]. By introducing the idea of interval tree, we can make a continuous model of binary search tree which is an important structure in Computer Science [Sibuya and Itoh (1987)]. The tree generated by the sequential bisection has less discrete structure than the original binary search tree and we can make an analogous analysis to the 1-dimensional random packing.

In general d-dimensional random cube packing problem are very hard to analyze mathematically for $d > 1$. Thus simulations are of paramount importance for the understanding of the involved phenomena. We give the general tools for doing random sequential packing by introducing the covering problem, Voronoi diagrams and Delaunay tessellations, which have a lot of applications in Science and Technology. We explain how this works out for the examples of packing by spheres, cubes and cross polytope (unit ball for the Hamming distance in Coding Theory).

We consider the simplest random sequential packing with rigid boundary, i.e. a packing in which cubes of sidelength 2 are put sequentially at random into the cube of sidelength 4, with a cubic grid of unit sidelength [Itoh and Ueda (1983)]. This simplest random sequential packing model seems to have some of the characteristics of the model by Palásti and to be related to the discrete geometry of cubes which has a long history of research going back to Minkowski. Computer Simulations up to dimension 11, suggest that the packing density γ_d satisfies $\gamma_d \simeq d^{-\alpha}$ with an appropriate constant α, as in the case of random sequential coding by Hamming distance [Itoh and Solomon (1986)]. Despite being in many ways a much simpler model, this model is still essentially unsolved. However by using the special geometry of the face lattice of the cube, it is shown that the expected number of decrease of the packing density is less than $\left(\frac{4}{3}\right)^d$ at each step of the random sequential packing. This shows that the expected number of cubes at the saturation is larger than $\left(\frac{3}{2}\right)^d$ [Poyarkov (2005)]. We also consider an extension of this model where one packs cubes of size N in a cube of side lengths $2N$.

This packing led us to the introduction of the random sequential packing of cubes of sidelength 2 in a torus of sidelength 4. Just like the previously considered rigid boundary model, this case allows us to introduce a combinatorial formalism to describe the corresponding cube packings. Thus we are able to apply the methods of exhaustive combinatorial enumeration to this setting. This allows us to find a number of remarkable combinatorial structures that could hardly have been found purely by hand computations. Then we consider a continuous variant of this model and we show that again we can describe it by a combinatorial model [Dutour Sikirić and Itoh (2010)]. This allows us to find some new remarkable structures and to find some extremal result on the number of parameters.

Whenever possible, we tried to give a description of the mathematical theory used when they are slightly non-standard and we give appendices on Complex Analysis, Laplace Transform, Renewal Theory, Space Groups and Exhaustive Combinatorial Enumeration. When several methods are possible to solve a problem, for example Tauberian Theorem and Complex Analysis, we have indicated both and discussed the relative merit of the possible approaches.

For discussion and suggestion on random sequential packing, we greatly thank J. E. Cohen, M. Deza, N. Dolbilin, M. Dutour Sikirić, J. M. Hammersley, T. Hattori, I. Higuti, F. Komaki, J. A. Morrison, Y. Nishiyama, A. Poyarkov, L. A. Shepp, R. Shimizu, M. Sibuya, H. Solomon, M. Tanemura, H. Tong and S. Ueda.

We thank also École Normale Supérieure of Paris, Institute of Statistical Mathematics, Hayama Center of the Graduate University for Advanced Studies, The Rockefeller University, National University of Ireland in Galway, Rudjer Bošković Institute of Zagreb, Technical University Delft, Mathematisches Forschungsinstitut Oberwolfach, Hausdorff Institute of Bonn for continued support.

Mathieu Dutour Sikirić and Yoshiaki Itoh

Contents

Chapter 1

Introduction

The classical packing problem is the sphere packing problem, i.e. the problem of putting balls of the same radius in Euclidean space $\mathbb{R}^n$ and maximize its density, i.e. minimize the volume of the interstitial space. The 2-dimensional sphere packing problem is solved by the hexagonal sphere packing [Lagrange (1773); Thue (1910); Zong (1999)]:

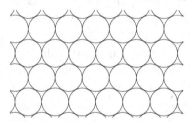

Despite having an *a priori* obvious solution with the stacking of hexagonal layers, the 3-dimensional sphere packing problem, also known as Kepler conjecture has resisted attempts for a long time until a solution was proposed in [Hales (2000, 2005)]. This solution was obtained by some computer computations and at the time of this writing the case of dimension 4 and higher is open. The density of a periodic cube packing of $\mathbb{R}^n$ is obtained by taking the volume covered by balls in a fundamental domain $\mathcal{F}$ divided by the volume of $\mathcal{F}$. If the structure is not periodic then it is possible to consider some cube $[-X, X]^n$ and to consider the limit as $X \to \infty$. Note that this limit does not necessarily exist and that the situation is even more complicated for spaces like hyperbolic space (see, for example, [Tóth and Kuperberg (1993); Tóth, Kuperberg and Kuperberg (1998)]). Partially because of those technical difficulties some restricted variants of the sphere packing problem have been introduced. One direction of research is finite packings [Böröcsky (2004)], where a finite region is considered and

1

some interesting phase transition phenomena occur [Zong (1999)]. Another direction of research is the lattice sphere packing problem, where the centers of the spheres are assumed to lie in a lattice, that is a subgroup $\mathbb{Z}v_1 + \cdots + \mathbb{Z}v_n \subset \mathbb{R}^n$. This subcase allows to use some more algebraic tools and the solution to the lattice sphere packing problem is known for dimension $n \leq 8$ and $n = 24$ (see, for example, [Martinet (2003); Zong (1999); Schürmann (2009)] for some exposition). See [Aste and Weaire (2008)] for more on packing theory.

A natural generalization of the sphere packing problem is to consider packing for a distance d, in the space $\mathbb{R}^n$. For a distance d on $\mathbb{R}^n$ denote by $B_d(x) = \{y \in \mathbb{R}^n \text{ s.t. } d(x,y) \leq 1\}$ the unit ball around x. We say that a family X of points defines a packings if the relative interiors of the corresponding balls do not self-intersect. Denote by $d_p(x,y) = \sqrt[p]{|x_1 - y_1|^p + \cdots + |x_n - y_n|^p}$ the L^p distance on $\mathbb{R}^n$. The distance d_2 is the Euclidean distance, the distance d_1 is sometimes called *Hamming Distance* in coding theory, see Chapter 9 for more details. For the distance d_1 the unit ball is a polytope named *cross polytope* with $2n$ vertices $x \pm e_i$, $1 \leq i \leq n$. The limit of d_p when $p \to \infty$ is the distance $d_\infty(x,y) = \max_i |x_i - y_i|$. The unit ball $B_{d_\infty}(x)$ is the cube of side length 2 centered at x. This book will be concerned with packing by cubes, that is for the d_∞ distance except for Chapter 9. Of course the optimal packing density for cubes is 1 since one can put them in such a way that they form a tiling of $\mathbb{R}^n$. More generally, any face-to-face tiling by cubes is the trivial one. Thus an interesting direction of research has considered non face-to-face tilings by cubes [Grünbaum and Shepard (1987); Bölcskei (2000); Martini (1998); Bölcskei (2001)]. Some applications to Coding Theory are considered in [Lagarias and Shor (1992, 1994)]. An interesting monograph on algebra and cube tilings is [Stein and Szabó (1994)]. Some relation of cube packings with Harmonic Analysis have been found [Zong (2005); Lagarias, Reeds and Wang (2000)] and some applications to complexity of algorithms even to statistics and music theory [Andreatta (2006)].

Here we will consider packings and sometimes cube packing by cubes with sides all of the same length. Such a packing is called non-extensible if one cannot add another cube to it.

One possible, perhaps the most simple, way to get a non-extensible packing in a set $X \subset \mathbb{R}^n$ is to add translates $B_d(x)$ at random in X until one cannot add any more such balls. This is the sequential random packing process that will occupy us most of the time, and which is a random cube packing process. If the vector x is selected with uniform probability among all possible vectors then we can consider the density of the obtained random packing as a random variable and study its mean, variance, etc. Quite often the variance turns out to be small, even negligible as the size of the domain increases to infinity. This explains why it makes sense to use such random sequential packing process in Statistical Mechanics, where one expects some form of isotropy and translational invariance to hold. Thus sequential random packings were applied to the modelization of disordered phenomena in physics; it was applied to discuss the geometric structure of liquids in [Bernal (1959)]. It has been further discussed in [Higuti (1960); Solomon (1967); Tanemura (1979); Dolby and Solomon (1975); Cohen and Reiss (1963)] and others for application to recognition and biological problems. More generally Statistical Mechanics on lattices is a huge field of research and one may consult, for example, [Evans (1993)].

In Figure 1.1 we show the result of sequential random packing for squares and circles. The problem is to consider the average of the random variable of the obtained packing density. The origin of the problem can be traced to children randomly putting balls in a box. Finding the limit density for this problem and actually any problem of random sequential packing is extremely difficult. Another problem is that physical situation may require another notion than sequential random cube packing since one may want to impose that no local improvement to the packing is possible. Imposing such kind of condition turns out to be harder than expected [Torquato (2000)]. Using the definition in [Torquato (2000)], recent work in theoretical physics [Song et al. (2008)] have indicated that the limit random packing density of sphere is 0.634.

The Flory model [Flory (1939)] consists of adding integral translates $x + [0, 2]$ with $x \in \mathbb{Z}$ into an interval $[0, n]$ until one cannot add any more intervals. This model was introduced in Polymer Chemistry in order to model the fraction of catalyst that did not react. This model admits generalization as integral translates of $[0, d]$ into $[0, n]$ and the limit as n and d go to infinity adequately is Rényi's random cube packing process.

We will consider sequential packing of cubes with particular attention to 1-dimensional problems. A 1-dimensional problem is the parking problem in a parking lane: put cars of unit length at random in an interval $[0, x]$ until it is not possible to add any more cars. The first recorded occurrence of this model is in [Rényi (1958)] and a variant is the energy cascade: the energy of a photon is split until it is no longer possible to do so [Ney (1962)]. Yet another problem is the random packing problem for elections as introduced

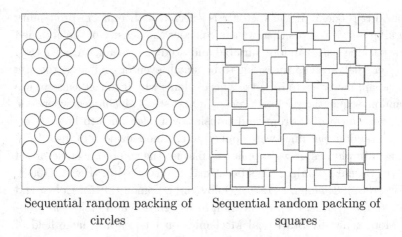

Sequential random packing of	Sequential random packing of
circles	squares

Fig. 1.1   Two 2-dimensional random sequential packing of squares and circles in a square box

in [Itoh and Ueda (1978, 1979); Itoh (1978, 1985)], to explain the territory making behavior of candidates. Consider a stick of length $x \geq 2d$. The stick is divided into two sticks with lengths x_1 and x_2 such that $x_1 \geq d$ and $x_2 \geq d$. Each possible division is assumed to be equally probable. Such division is continued until all sticks are shorter than $2d$. The sticks obtained by such a procedure correspond to gaps generated by 1-dimensional random packing. In the model, the length of each stick corresponds to the votes obtained by a candidate. Other biological applications are considered, for example in [Tanemura and Hasegawa (1980)]. Another application of sphere packing is to Recognition Theory [Dolby and Solomon (1975)].

[Palásti (1960)] considered non-extensible packings obtained from sequential random packings of cubes $[0, 1]^n$ into $[0, x]^n$. She conjectured that the expectation $E(M_n(x))$ of the packing density $M_n(x)$ satisfies the limit

$$\lim_{x \to \infty} \frac{E(M_n(x))}{x^n} = \beta_n \qquad (1.1)$$

with $\beta_n = \beta_1^n$. Several simulations and analysis in [Blaisdell and Solomon (1970, 1982); Akeda and Hori (1975, 1976)] makes the conjecture very unlikely, although it has not been formally disproved. Moreover, based on those simulations it is expected that $\beta_n > \beta_1^n$ and an experimental formula

$$\beta_n^{1/n} - \beta_1 \simeq (n-1)(\beta_2^{1/2} - \beta_1)$$

has been proposed [Blaisdell and Solomon (1982)] from those simulations. Alfred Rényi (see [Rényi (1958)]) has proved that

$$C_R = \beta_1 = \int_0^\infty \exp\left\{ -2 \int_0^t \frac{1 - \exp(-u)}{u} du \right\} dt.$$

The value of this integral was later estimated to be about 0.7475979202 [Blaisdell and Solomon (1970)]. Rényi's exact analysis for the random parking of unit length cars is based on the observation that after the first car parks on a finite interval, one is left with car parking problems on two smaller intervals. This idea, commonly used in 1-dimensional problems, is also used in Flory's combinatorial argument [Flory (1939)] for random sequential dimer filling. Rényi's method to get the parking constant is based on the Laplace transform of the expected number of cars and can be applied to various situation of random space filling; it is presented in Chapter 3. Another generalization has been considered in [Ney (1962)] where the length itself is a random variable. Again the solution method is Laplace transform of the expected packing density. Higher dimensional problems are considerably harder and the only known result is the proof of the existence of the limit (1.1) by [Penrose (2001)] (see [Penrose and Yukich (2001, 2002, 2003); Schreiber, Penrose and Yukich (2007); Baryschnikov and Yukich (2002)] for some recent developments in this line of research).

After Rényi's initial work, a number of results were proved. [Dvoretzky and Robbins (1964)] proved the convergence by another, much simpler technique, and this allowed them to give a very good estimate on the rate of convergence of

$$\frac{E(M_1(x))}{x}$$

to C_R. They showed that it converges faster than the exponential and that the variance grows linearly. They proved a central limit theorem for Rényi's model (Chapter 5). Further analytical studies of the minimum of gap was considered in [Itoh (1980)] (Chapter 4). By using Renewal Theory from classical probability theory, it was shown that the probability that all gaps are larger than h decreases exponentially with $a(h)$ being the exponent of decrease. The quantity $a(h)$ is obtained as the explosion time of an ordinary differential equation and detailed asymptotic analysis of $a(h)$ were carried out as h goes to 0 and 1 by analyzing the singular perturbation problems.

In Chapter 6 we consider the Kakutani interval splitting model, where one splits $[0, 1]$ at random and continues to split the interval of largest length. We study this model by associating to any such sequence a binary tree [Sibuya and Itoh (1987)]. The structure of binary tree is a fundamental building block of Computer Science and the theory of random trees is fundamental to the analysis of algorithms [Drmota (2009); Mahmoud (1992); Flajolet and Sedgewick (2009)]. We are able to prove that this tree has the same asymptotic behavior as the random binary tree obtained by the quicksort algorithm. Also, using the tree structure we can give a simple proof of the uniform distribution of the size of gaps.

In Chapter 7 we consider the unified Kakutani Rényi model, where cars of length l are put in an interval of length x when the gap is not less than 1 [Komaki and Itoh (1992)]. For this model, we proved the finiteness of the moment of the number of interval packed (a non-obvious result if $l = 0$) and a strong law of large number. Then we proved a central limit theorem and we determine the distribution of the average size of gaps. If $l = 0$, then the model can be reinterpreted (see [Lootgieter (1977); Pyke (1980); Van Zwet (1978)]) as Kakutani's interval splitting model. We can prove more for Kakutani's interval splitting model: we can determine its mean and variance explicitly and we can prove that the length of gaps is uniformly distributed on $[0, 1]$. We also determine the distribution of a gap chosen at random, by following the argument of [Bankövi (1962)] and adapting them to this case.

Then we consider the problem of parking with spin where cars have two ways to park (Chapter 8). We consider cars of length 0 which can park up with probability p and down with probability $1 - p$ [Itoh and Shepp (1999)]. They should be at distance at least 1 from a car of the same spin and distance at least a from a car of opposite spin. Actually, the theory is quite similar to the one of Rényi's problem except that one has to consider three different functions and a linear matrix ordinary differential equation of first order for them. We proved the existence of the limit packing density by using a diagonalization into eigenvectors and using Dvoretzky-Robbins theory as well as more simpler integral equations arguments. We found a number of exactly solved cases for which we can give an explicit integral expression of the random packing density. But in general it is not possible to solve the system of integral equations and one has to be content with numerical solutions. For that purpose a power series solution is found and the obtained results show an interesting non-monotonicity. We should point out that many variants of this problem are possible for 1-dimensional problems.

Since analytic methods have very limited success for dimension higher than 2, we have to use some simulations for obtaining if not rigorous results at least some ideas about what they should be. In order to simulate sequential random cube packing, one has to choose at random the cubes that one adds and to test whether or not a cube packing is extensible. Testing extensibility is actually equivalent to determining the covering radius of a point set. So, we introduce the notions of Voronoi diagram, Delaunay cells and then we explain the known results for best packing and random sequential packing of spheres, cubes and cross polytopes. Finally, we give some simulation results for the Golay code.

In Chapter 10 we consider the model [Itoh and Ueda (1983)], which is perhaps the simplest high dimensional random sequential cube packing problem: It considers packing of cubes $z + [0, 2]^n$ with $z \in \mathbb{Z}^n$ into $[0, 4]^n$. Simulations for this model indicate that

$$E(M_2^C(n)) \sim \frac{2^n}{d^\alpha}$$

for some empirical constant α [Itoh and Solomon (1986)]. It is proved in [Poyarkov (2005, 2003)] that the average number of cubes satisfies the inequality $E(M_2^C(n)) \geq \left(\frac{3}{2}\right)^n$, which is the first non-trivial lower bound on the packing density. The method is to notice that cube packing of $[0, 2]^n$ in $[0, 4]^n$ can actually be encoded by the face lattice of the cube.

In Chapter 11 we consider cube packings of $\mathbb{R}^n$ by cubes $z + [0, 2]^n$ with $z \in \mathbb{Z}^n$ that are periodic along the lattice $4\mathbb{Z}^n$ [Dutour Sikirić, Itoh and Poyarkov (2007)]. As for the case of the model by Itoh and Ueda, we can encode such cube packing in a neat way with a graph. We are thus able to apply techniques from exhaustive combinatorial enumeration, which we briefly survey in Appendix A. We enumerate completely the non-extensible cube packings in dimension 3 and 4 and this allows us to find a number of interesting remarkable structure of high symmetry. One of them is a packing with cubic symmetry of $\mathbb{R}^3$ that is non-extensible and of half density. This cube packing, which was first introduced in [Lagarias, Reeds and Wang (2000)] is the maximal type of non-extensible cube packing, which is not a cube tiling. For the minimal number $f(n)$ of cube of non-extensible cube packing, we found $f(3) = 4$, $f(4) = 8$, $10 \leq f(5) \leq 12$ and $14 \leq f(6) \leq 16$. We also obtain some lower and upper bounds for the second moment of the number of cubes contained in a translate of a cube $[0, 4]^n$.

In Chapter 12 we consider packing of cubes $[0, 1]^n$ which are periodic along $2\mathbb{Z}^n$ but we no longer impose that the cubes are integral translates [Dutour Sikirić and Itoh (2010)]. But due to the fact that the cube have length 1 in a space of length 2 the packing condition is expressed simply as $|z_i - z_i'| = 1$ and this allows to introduce a combinatorial formalism for dealing with them. For example the limit packing density is always rational and the corresponding cube packings are described in simple way in terms of some parameters. We prove that in odd dimension the minimal number of cube in a non-extensible cube packing is $n + 1$ and is described by a 1-factorization of the complete graph K_{n+1}. A very interesting 3-dimensional cube tiling is the rod tiling where infinite lines of cubes go in three different directions and two translation classes of cubes fill the gap. We generalized this structure to dimension 5, 7 and 9 but not 11 for unclear reasons. We also conjectured that the number of parameters describing such continuous cube packings is at most $2^n - 1$. Much remains to be done in this nascent theory.

Packing problems are merely the most simple discrete model in the discrete geometry of point sets. The next most important problem is the covering problem, where one uses spheres (or cubes, etc.) and impose the condition that every point belongs to at least one of those spheres. Covering problems are generally more complicated than packing problems as the monograph [Schürmann (2009)] explains nicely in the case of lattices. There is also a theory of random sequential coverings. The basic stochastic process is that one puts intervals of length l_1, l_2, ..., l_n with $l_i < 1$ at random on a circle of length 1. The set of all those intervals covers the circle except possibly a set of measure 0 with probability 1 if and only if $\sum_{i=1}^{\infty} l_i = \infty$. [Shepp (1972)] proved that those intervals cover the whole circle if and only if

$$\sum_{n=1}^{\infty} \frac{1}{n^2} \exp(l_1 + \cdots + l_n) = \infty.$$

This result generalizes previous criterions of [Dvoretzky (1956); Mandelbrot (1972); Flatto and Konheim (1962)]. This is generalized to the line in [Shepp (1972)], and to higher dimensional cases in [Kahane (1985)].

Chapter 2 considers the problem of packing intervals $[0, d]$ in $[0, n]$ and computes the packing density for the case $d = 2$ of Flory model. This chapter uses less technicalities than the continuous case of Rényi's model and is not used later on. Chapters 3 to 8 are all related and are best read sequentially. Chapters 9 to 12 can all be read independently.

Chapter 2

The Flory model

We consider here a discrete sequential random packing process where we pack intervals $[x, x + d]$ with $x \in \mathbb{Z}$ in the interval $[0, n]$ at random. It is a discrete version of Rényi's problem considered in Chapter 3. Thus it is simpler to study asymptotically but the obtained results are less precise than for the continuous version. The case $d = 2$ is named Flory's model after [Flory (1939)] introduced it to treat some problems in catalytic chemistry (see Figure 2.1 for an example).

We study the existence of the limit packing density and the distribution of the number of gaps. The method is to write down some recurrence equations and to study them by recursion arguments or by using generating functions.

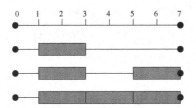

Fig. 2.1   Flory cube packing in the interval $[0, 7]$

2.1   One-dimensional discrete random sequential packing

Place at random an interval $I_1 = [\xi, \xi + d]$ of length d in $[0, n]$. The initial point ξ of the interval I_1 is a uniformly distributed random variable. The left edge ξ of the interval I_1 is placed uniformly at random from

$\{0, 1, 2, \cdots, n - d\}$. Then place another interval of length d at random, independently of the location of the interval I_1, to the interval $[0, n]$. If the second interval does not intersect I_1, we keep it and denote it by I_2; if it does, we neglect it and choose a new interval. If the (disjoint) intervals $I_1, I_2, \ldots, I_k$ have already been chosen, the next randomly chosen interval will be kept only if it does not intersect any of the intervals $I_1, I_2, \ldots, I_k$. In this case this interval will be denoted by I_{k+1}. The procedure is continued as long as possible, i.e. until the following situation occurs: none of the distances between consecutive intervals is greater than $d - 1$ and the endpoints of the interval $[0, n]$ are near to an interval within distance less than d. Assume that this occurs after placing $v_d(n)$ intervals. The problem is to determine the expectation $M_d(n) = E(v_d(n))$ and the limit

$$\lim_{n \to \infty} \frac{M_d(n)}{n} = C_d.$$

Obviously

$$M_d(n) = \begin{cases} 0 & \text{if } 0 \le n \le d - 1, \\ 1 & \text{if } d \le n < 2d. \end{cases}$$

It is clear that $0 \le M_d(n) \le n/d$.

If an interval is already placed on the interval $[0, n + d]$ and it is the interval $[t, t + d]$ for $0 \le t \le x$, then the average number of intervals on the left is $M_d(t)$ and that of intervals on the right is $M_d(n - t)$, and thus since t is uniformly distributed on $\{0, 1, 2, \cdots, n\}$ we have

$$\begin{aligned} M_d(n + d) &= 1 + \frac{1}{n+1} \sum_{t=0}^{n} M_d(t) + M_d(n - t) \\ &= 1 + \frac{2}{n+1} \sum_{t=0}^{n} M_d(t). \end{aligned} \tag{2.1}$$

2.2 Application of generating function

In order to deal with recursion formula it is very convenient to use formal power series. This technique dating back to DeMoivre and Laplace allows for very powerful analytical tools to be used.

Definition 2.1. Let $a_0, a_1, a_2, \ldots$ be a sequence of real numbers. The series

$$A(s) = a_0 + a_1 s + a_2 s^2 + \ldots \tag{2.2}$$

is called the *generating function* of the sequence (a_j).

There exists a radius of convergence $r \in \mathbb{R}_+ \cup \{\infty\}$ such that the series $\sum_{i=0}^{\infty} a_i s^i$ converge for $|s| < r$ and diverge for $|s| > r$ with the behavior for $|s| = r$ varying from case to case. The radius of convergence of a generating function is allowed to be 0. For example, the generating function $\sum_n n! x^n$ is allowed. We can define the differentiation, integration and product of generating functions simply by considering the objects to be formal and doing operation term by term. But in all examples that we will consider the generating functions will have a positive radius of convergence. For example if the sequence (a_j) is bounded, then a comparison with the geometric series shows that (2.2) converges at least for $|s| < 1$. The variable s itself has no signification *a priori*. Inside the disk of convergence it is possible to differentiate term by term the power series and thus translate recursion relations into differential or functional equations. The interested reader can consult [Wilf (2006); Graham, Knuth and Patashnik (1994); Flajolet and Sedgewick (2009)] for further exposition of the theory.

Take for example the sequence

$$x_k = \begin{cases} 0 \text{ if } k \leq m - 1, \\ 1 \text{ if } k \geq m. \end{cases}$$

Then the corresponding generating function is

$$f(s) = \sum_{k=m}^{\infty} s^k = s^m \frac{1}{1-s}.$$

If $x_k = \frac{1}{k!}$ then the generating function is e^x. If x_k is zero for large values of k then the generating function is a polynomial. Generating functions are useful because they allow the use of very efficient techniques of power series, differential equations and complex analysis.

Let us write the generating function

$$f_d(s) = \sum_{n=d+1}^{\infty} M_d(n) s^n.$$

Theorem 2.1. *(i) The generating function $f_d(s)$ converges for $|s| < 1$.*
(ii) It satisfies the differential equation

$$f_d'(s) = \left[\frac{2s^{d-1}}{1-s} + \frac{d-1}{s} \right] f_d(s) + \frac{2s^{2d-1}}{1-s} + \frac{s^d}{1-s} + \frac{s^d}{(1-s)^2}. \tag{2.3}$$

(iii) If $d = 2$ then we have the following expression:

$$f_2(s) = \frac{s}{2(1-s)^2} - s^2 - \frac{1}{2} \frac{e^{-2s} s}{(1-s)^2}.$$

Proof. We know that $0 \leq M_d(n) \leq n$ which implies that the generating function $f_d(s)$ actually converges for $|s| < 1$ and thus (i).

Equation (2.1) is rewritten as

$$(n+1)M_d(n+d) = n+1+2\sum_{t=0}^{n} M_d(t),$$

which implies by subtraction

$$(n+1)M_d(n+d) - nM_d(n+d-1) = 1 + 2M_d(n).$$

We have the summation

$$
\begin{aligned}
\sum_{n=1}^{\infty}(n+1)M_d(n+d)s^n &= \left(\sum_{n=1}^{\infty} M_d(n+d)s^{n+1}\right)' \\
&= \left(\sum_{n=d+1}^{\infty} M_d(n)s^{n+1-d}\right)' \\
&= (s^{1-d}f_d(s))'
\end{aligned}
$$

and we have similarly

$$
\begin{aligned}
\sum_{n=1}^{\infty} nM_d(n+d-1)s^n &= s\sum_{n=0}^{\infty}(n+1)M_d(n+d)s^n \\
&= s(s^{1-d}f_d(s))' + s.
\end{aligned}
$$

Finally we have

$$\sum_{n=1}^{\infty}\{1+2M_d(n)\}s^n = \frac{s}{1-s} + 2f_d(s) + 2s^d,$$

which implies the differential equation

$$(s^{1-d}f_d(s))' - s(s^{1-d}f_d(s))' - s = \frac{s}{1-s} + 2f_d(s) + 2s^d.$$

This equation is rewritten as (2.3) in (ii).

Let us consider the case $d = 2$. The solution of the homogeneous equation

$$f_2'(s) = \left[\frac{2s}{1-s} + \frac{1}{s}\right] f_2(s)$$

is

$$f_{2,0}(s) = \frac{e^{-2s}s}{(1-s)^2}.$$

If one writes $f_2(s) = f_{2,0}(s)u(s)$ (hence the name of the method "variation of constant") then Equation (2.3) is rewritten as the equation

$$u'(s) = (2s + s^2 - s^3)e^{2s}$$

whose integration gives

$$u(s) = (1/2 - s + 2s^2 - s^3)e^{2s} + c_1.$$

Since $f_2'(0) = 0$, we have $c_1 = -1/2$ and thus finally:

$$f_2(s) = \frac{s}{2(1-s)^2} - s^2 - \frac{1}{2} \frac{e^{-2s}s}{(1-s)^2} \tag{2.4}$$

that is the expression in (iii). □

Theorem 2.2. *For $d = 2$ we have the asymptotic expansion*

$$M_2(n) = \frac{1}{2}(1 - (1/e^2))n - e^{-2} + O\left(\frac{1}{a^n}\right) \tag{2.5}$$

for all $a > 1$.

Proof. The function $f_2(s)$ was originally defined over the disk of convergence $D = \{z \in \mathbb{C} : |z| < 1\}$. But the expression from (2.4) shows that it can be extended to $\mathbb{C} - \{1\}$. Now if one uses the Taylor expansion of the function e^{-2s} around $s = 1$ and simplify one gets:

$$f_2(s) = \frac{1}{(1-s)^2} \frac{1 - e^{-2}}{2} - \frac{1}{1-s} \frac{1 + e^{-2}}{2} + \psi(s)$$

with $\psi(s)$ a function entire over $\mathbb{C}$. Note that this kind of expansion around a singularity z_0 in powers of $(z - z_0)^i$ and $\frac{1}{(z-z_0)^i}$ are called Laurent series. Let us denote by x_k the coefficients of the power series of $\psi(s)$. Since the power series of ψ converges on $\mathbb{C}$, the radius of convergence is infinite. This implies $x_k a^k$ converging to zero for all $a > 0$ and thus $x_k = O\left(\frac{1}{a^n}\right)$. We have the following expansions:

$$\frac{1}{(1-s)^2} = \sum_{k=0}^{\infty}(k+1)s^k \text{ and } \frac{1}{1-s} = \sum_{k=0}^{\infty} s^k.$$

Equation (2.5) follows by combining all previous expansions. □

It is possible to get the first order by applying a Tauberian theorem [Feller (1971)] but Tauberian theorem techniques have the problem of being more complicated to prove, of depending on positivity hypothesis and of being difficult to use for higher order.

2.3 Number of gaps

The direct approach is useful to make numerical studies to understand the behavior of the random sequential packing. Let us study the argument by [Flory (1939)]. Assume that the probability of reaction of each adjacent pair of unreacted components is the same and independent of the status of neighboring components, the fraction of the X's which become isolated, and therefore remain unreacted at the termination of the reaction, can be calculated in the following manner. Consider a group of polymer molecules each containing n structural units (and n substituents) in the 1, 3 configuration (1). The average number of unreacted $X's$ per molecule at the end of reaction will be termed $S_2(n)$. Obviously $S_2(0) = 0$, $S_2(1) = 1$ and $S_2(2) = 0$. In $n = 4$, the first reaction may link any one of three pairs, which may be designated 1-2, 2-3, 3-4. If either the 1-2 or the 3-4 pair is joined, the unreacted portion is equivalent to an $n = 2$ molecule; if the 2-3 pairs is joined, there remains the equivalent of two $n = 1$ molecules. Hence

$$S_2(4) = (2S_2(0) + 2S_2(1) + 2S_2(2))/3.$$

If $n = 5$ the first reaction may join any one of four different pairs. Two of these yield an $n = 3$ remainder; each of the other two possibilities yields both $n = 2$ and $n = 1$ remainder. Hence

$$S_2(5) = (2S_2(0) + 2S_2(1) + 2S_2(2) + 2S_2(3))/4.$$

Continuing in this manner, we obtain the following result:

Theorem 2.3. *We have for $n \geq 3$ the following equality:*

$$S_2(n) = \frac{2}{n-1}(S_2(0) + S_2(1) + S_2(2) + \cdots + S_2(n-2)). \qquad (2.6)$$

We will now use the above recursive result to derive in an elementary way, that is without generating functions Flory's result:

$$S_2(n) \simeq n/e^2.$$

So, the fraction of substituents which become isolated is very nearly $1/e^2 = 0.1353\ldots$, or 13.53 percent.

Theorem 2.4. *(i) We have*

$$S_2(n) \simeq n/e^2.$$

(ii) We have the explicit expression

$$S_2(n) = \sum_{k=0}^{n-1}(n-k)\frac{(-2)^k}{k!}.$$

Proof. Subtracting the expression for $(n-2)\,S_2(n-1)$ from the expression for $(n-1)\,S_2(n)$ and setting $\triangle_n = S_2(n) - S_2(n-1)$, we have

$$\begin{cases} (n-1)\triangle_n + S_2(n-1) = 2S_2(n-2), \\ (n-2)\triangle_{n-1} + S_2(n-2) = 2S_2(n-3). \end{cases}$$

Considering $(n-1)\triangle_n = (n-2)\triangle_n + \triangle_n$, we have

$$\triangle_n - \triangle_{n-1} = \frac{-2}{n-1}(\triangle_{n-1} - \triangle_{n-2})$$

or

$$\triangle_n - \triangle_{n-1} = \frac{(-2)^{n-1}}{(n-1)!}(\triangle_1 - \triangle_0).$$

Recalling the value of $S_2(0)$, $S_2(1)$ and $S_2(2)$, $\triangle_1 = 1$ and $\triangle_2 = -1$. From (4)

$$\triangle_2 - \triangle_1 = -2(\triangle_1 - \triangle_0) = -2$$

and therefore $\triangle_1 - \triangle_0$ must equal unity, and by simple summation we get

$$\triangle_n = 1 - \frac{(2)}{1!} + \frac{4}{2!} - \frac{8}{3!} + \cdots + \frac{(-2)^{n-1}}{(n-1)!}.$$

This sequence is the partial sum of the power series of the exponential function evaluated at -2. It is alternating and it converges very fast when $n \to \infty$; i.e.

$$\left|\triangle_n - \frac{1}{e^2}\right| \le \frac{2^n}{n!} \quad \text{and} \quad \lim_{n\to\infty} \triangle_n = \frac{1}{e^2}. \tag{2.7}$$

We remind the reader of Cesaro's theorem that states that if a sequence u_n converges towards a limit l then the sequence

$$\frac{u_0 + \cdots + u_n}{n+1}$$

converges towards l as well. For $S_2(n)$ Cesaro's theorem combined with (2.7) implies:

$$S_2(n) \simeq n/e^2$$

that is (i).

When the molecule have short length it might be advantageous to have an explicit expression for $S_2(n)$. The following expression is easily computed:

$$\begin{aligned} S_2(n) &= S_2(0) + \sum_{k=1}^{n} S_2(k) - S_2(k-1) \\ &= \sum_{k=1}^{n} \triangle_k \\ &= \sum_{k=1}^{n} \sum_{l=0}^{k-1} \frac{(-2)^l}{l!} \\ &= \sum_{k=0}^{n-1}(n-k)\frac{(-2)^k}{k!} \end{aligned}$$

that is (ii) holds.　　　　　　　　　　　　　　　　　　　　　□

For the general case let $S_d(n)$ denote the number of gaps for the random packing. By the same technique as for the case $d = 2$ we can prove that $S_d(n)$ follows the following recursive relation:

$$S_d(n) = \frac{2}{n-d+1}(S_d(0) + S_d(1) + S_d(2) + \cdots + S_d(n-d)),$$

and we define the generating function

$$g_d(s) = \sum_{n=d+1}^{\infty} S_d(n)s^n.$$

Theorem 2.5. *(i) The generating function g_d converge for $|s| < 1$.*

(ii) The function g_d satisfies the differential equation

$$g_d'(s) = g_d(s)\frac{-1+d-(d-1)s+2s^d}{s-s^2} + 2s^d\frac{1}{s-s^2}\sum_{k=1}^{d-1}s^k.$$

(iii) If $d = 2$ then

$$g_2(s) = -s + \frac{e^{-2s}s}{(1-s)^2}. \tag{2.8}$$

Proof.　For $n \geq d + 1$, considering the argument on (2.6) we have

$$(n-d+1)S_d(n) - (n-d)S_d(n-1) = 2S_d(n-d).$$

The generating function $g_d(s)$ converges for $|s| < 1$ since we have the inequality $0 \leq S_d(n) \leq n$ on the number of gaps. Consider

$$\sum_{n=d+1}^{\infty}(n-d+1)S_d(n)s^n - \sum_{n=d+1}^{\infty}(n-d)S_d(n-1)s^n = \sum_{n=d+1}^{\infty} 2S_d(n-d)s^n.$$

By the above argument we have

$$2s^d(\textstyle\sum_{k=1}^{d}s^kS_d(k)) + S_d(d)s^{d+1} = (1-d+(d-1)s-2s^d)g_d(s) + (s-s^2)g_d'(s)$$

and

$$g_d'(s) = g_d(s)\frac{-1+d-(d-1)s+2s^d}{s-s^2}$$
$$+ (2s^d(\textstyle\sum_{k=1}^{d}s^kS_d(k)) + S_d(d)s^{d+1})\frac{1}{s-s^2}.$$

Since $S_d(0) = 0$, $S_d(k) = 1$ for $k = 1, 2, \cdots, d-1$ and $S_d(d) = 0$, we have

$$g_d'(s) = g_d(s)\frac{-1+d-(d-1)s+2s^d}{s-s^2} + 2s^d\frac{1}{s-s^2}\sum_{k=1}^{d-1}s^k,$$

which is the required equation.

Consider Flory's model for the case $d = 2$. By applying the variation of constant method which we have already seen in Section 2.2, we get:

$$g_2(s) = -s + c_2 \frac{e^{-2s}s}{(-1+s)^2}.$$

Since $g_2'(0) = 0$, we have $c_2 = 1$ and the result follows.   □

Note that by applying the same technique as in Theorem 2.2, we can give another proof that

$$S_2(n) \simeq (1/e^2)n, \tag{2.9}$$

for sufficiently large n. This means that for a high molecular weight polymer where n is large the fraction of the substituents which becomes isolated is very nearly $1/e^2 = 0.1353\ldots$ as was shown by the argument by Flory.

Considering

$$2f_2(s) + g_2(s) = \sum_{k=d+1}^{\infty} ks^k, \tag{2.10}$$

(2.8) and (2.9) are also obtained from (2.4) and (2.5), since we see that $2M_2(n) + S_2(n) = n$. But there is no such simple formula for the case $d \geq 3$.

2.4   Minimum of gaps

Before explaining the general equation, let us introduce the notion of convolution for integer valued random variables. Suppose that X and Y are independent random variables with integral non-negative values, i.e. $Pr(X = k) = x_k$ and $Pr(Y = k) = y_k$.

Then we have by summing over all possibilities

$$
\begin{aligned}
Pr(X + Y = n) &= \sum_{k=0}^{n} Pr(X = k, Y = n - k) \\
&= \sum_{k=0}^{n} Pr(X = k)Pr(Y = n - k) \\
&= \sum_{k=0}^{n} x_k y_{n-k},
\end{aligned}
$$

where we have used the independence to compute $Pr(X = k, Y = n - k)$. The above transformation where one takes a series $x = (x_0, x_1, \ldots)$ and a series $y = (y_0, y_1, \ldots)$ and get another series $z = (z_0, z_1, \ldots)$ with

$$z_n = \sum_{k=0}^{n} x_k y_{n-k}$$

is called the convolution of the series x and y. Now if f, g and h are the generating functions of X, Y and $X + Y$ then we get

$$
\begin{aligned}
h(s) &= \sum_{n=0}^{\infty} z_n s^n \\
&= \sum_{n=0}^{\infty} \left\{ \sum_{k=0}^{n} x_k y_{n-k} \right\} s^n \\
&= \sum_{n=0}^{\infty} \left\{ \sum_{k=0}^{n} x_k s^k y_{n-k} s^{n-k} \right\} \\
&= \left\{ \sum_{k=0}^{\infty} x_k s^k \right\} \left\{ \sum_{k=0}^{\infty} y_k s^k \right\} \\
&= f(s)g(s).
\end{aligned}
$$

This is one example of how the generating functions allow easy translations of properties of series.

Consider the general case of discrete random sequential packing. The cars of length d park sequentially at random at a street $[0, n]$ of length n. The left edge of each car is placed uniformly at random from the points $0, 1, 2, \cdots, n - d$. Let the minimum of gaps generated by the cars parked at a street of length k be L_k, write $P_h(k) = Pr(h \le L_k)$ and

$$
l_h(s) = \sum_{n=d}^{\infty} P_h(n) s^k.
$$

We have the following theorem:

Theorem 2.6. *(i) The numbers $P_h(k)$ follow the following equation for $d \le n$:*

$$
P_h(n) = \frac{1}{n - d + 1} \sum_{k=0}^{n-d} P_h(n - d - k) P_h(k).
$$

(ii) The generating function $l_h(s)$ converges for $|s| < 1$.
(iii) l_h satisfies the following differential equation

$$
\frac{d}{ds} \left(s^{-(d-1)} l_h(s) \right) = \left(l_h(s) + \sum_{k=h}^{d-1} P_h(k) s^k \right)^2. \tag{2.11}
$$

(iv) If $h = d - 1$ then we have

$$
l_h(s) = \left(\frac{1}{1 - \frac{s^{2d-1}}{2d-1}} - 1 \right) s^{d-1}.
$$

Proof. Let us denote by k the position of the first interval. By conditioning we have the formula

$$
P_h(n) = \frac{1}{n - d + 1} \sum_{k=0}^{n-d} Pr(h \le L_n | k).
$$

The interval $[k, k+d]$ splits $[0, n]$ into two intervals of length k and $n-d-k$ and the condition on the minimal size of gaps must be satisfied on both intervals. Thus $Pr(h \leq L_n|k) = Pr(h \leq L_k)Pr(h \leq L_{n-d-k})$ and (i) follows.

The convergence of the generating function for $|s| < 1$ follows from the inequality $0 \leq P_h(k) \leq 1$ for each k.

Consider

$$\sum_{n=d}^{\infty}(n - d + 1)P_h(n)s^n = \sum_{n=d}^{\infty}\sum_{k=0}^{n-d} P_h(n - d - k)P_h(k)s^n.$$

We have

$$\begin{aligned}
\sum_{n=d}^{\infty}(n - d + 1)P_h(n)s^n &= s^d \sum_{n=d}^{\infty}(n - d + 1)P_h(n)s^{n-d} \\
&= s^d \sum_{n=d}^{\infty}\frac{d}{ds}P_h(n)s^{n-d+1} \\
&= s^d \frac{d}{ds} \sum_{n=d}^{\infty} P_h(n)s^{n-d+1} \\
&= s^d \frac{d}{ds} \left(s^{-(d-1)} \sum_{n=d}^{\infty} P_h(n)s^n\right) \\
&= s^d \frac{d}{ds} \left(s^{-(d-1)}l_h(s)\right).
\end{aligned}$$

We have

$$\begin{aligned}
&= \sum_{n=d}^{\infty}\sum_{k=0}^{n-d} P_h(n - d - k)P_h(k)s^n \\
&= \sum_{n=d}^{\infty}\sum_{k=0}^{n-d} P_h(n - d - k)s^{n-d-k}P_h(k)s^k s^d \\
&= \left(\sum_{k=0}^{\infty} P_h(k)s^k\right)^2 s^d \\
&= s^d \left(l_h(s) + \sum_{k=h}^{d-1} P_h(k)s^k\right)^2.
\end{aligned}$$

Hence we have

$$s^d \frac{d}{ds} \left(s^{-(d-1)}l_h(s)\right) = s^d \left(l_h(s) + \sum_{k=h}^{d-1} P_h(k)s^k\right)^2,$$

which is the required equation of (iii).

If $h = d - 1$ then we have

$$\frac{d}{ds} \left(s^{-(d-1)}l_h(s)\right) = s^{2(d-1)} \left(s^{-(d-1)}l_h(s) + 1\right)^2, \tag{2.12}$$

which gives

$$\frac{d}{ds} \frac{1}{s^{-(d-1)}l_h(s) + 1} = s^{2(d-1)}. \tag{2.13}$$

The solution is then obtained by simple integration. $\qquad \square$

Actually for the case $h = d - 1$ the direct strategy is simpler:

Proposition 2.1. *If $n = d - 1 + (2d - 1)k$, then we have*

$$P_{d-1}(n) = \frac{1}{(2d - 1)^k},$$

otherwise $P_{d-1}(n) = 0$.

Proof. We limit ourselves to the case $d = 2$ which encodes all the features of the problem. If $n \neq 1 + 3k$ for all k then necessarily two intervals $[x, x+2]$ and $[y, y+2]$ are consecutive and this implies that $P_0(n) = 1$. If $n = 1 + 3k$ then the only way to have $L_k \geq 1$ is that all the intervals $[x, x+2]$ are organized with $x \in \{1, 1+3, \ldots, 1+3(k-1)\}$. If we condition over the first interval then we get

$$P_1(n) = P_1(1 + 3k) = \frac{1}{3k} \sum_{l=0}^{k-1} P_1(1 + 3l) P_1(1 + 3(k - 1 - l)).$$

Thus the result follows by induction and the general case likewise. $\square$

The proposition gives the probability of getting the most sparse packing for the case $n = d - 1 + (2d - 1)k$. Put $\frac{n}{d} = x$, $P_{d-1}(n) = Q_h(x)$ with $\frac{d-1}{d} = h$, we have

$$\log Q_h(x) \sim -\frac{1}{2} \log \frac{2}{1 - h}(x - h),$$

which suggests the asymptotic behavior of the continuous case given in Subsection 4.4.1.

2.5   Packing on circle and numerical study

From numerical studies on the random packing it appears that the quantity

$$\frac{M_d(n)d + d}{n + d} \tag{2.14}$$

seems to give a rapid convergence towards a constant C_d for each d. Of course, C_d converges to Rényi's constant C_R as $d \to \infty$. For example we got 0.74803 for the case $d = 500$ and $n = 3000$. One of the possible reasons why formula (2.14) seems to converge faster to its limit is that it expresses the density of a random packing on the circle of length $n + d$ by intervals of length d. In particular for $d = 2$, formula (2.5) can be rewritten as

$$\frac{2M_2(n) + 2}{n + 2} = 1 - e^{-2} + O\left(\frac{1}{a^n}\right)$$

for all $a > 1$.

We will see this again in Section 3.4 and Chapter 5 for the continuous random sequential packing. This strategy is very efficient in computing packing density and thus getting estimates on the limit packing density for 1-dimensional problems; it is used for example in Chapter 8.

2.6 Appendix: Complex Analysis

We now explain some notions of Complex Analysis that we need for this chapter and several other chapters. There are literally thousands of books on Complex Analysis and most of them are good, any one of them will be enough for complement on the notions shortly explained here.

A function f is called complex if it is defined over an open connected subset $D \subset \mathbb{C}$, called domain, with values in $\mathbb{C}$. It is called differentiable at $a \in D$ if the limit

$$\lim_{z \to a} \frac{f(z) - f(a)}{z - a}$$

exists. The familiar differentiation rules of calculus also hold for complex functions. A complex function $f : D \to \mathbb{C}$ is called holomorphic, or analytic, if it is differentiable for any $a \in D$. Then, if $D(a, r)$ denotes the disk of center a and radius r and $D(a, r) \subset D$ then for any z with $|z - a| < r$ we have

$$f(z) = \sum_{k=0}^{\infty} \frac{(z - a)^k}{k!} f^{(k)}(a) \tag{2.15}$$

that is the Taylor expansion at a converges in the interior of the disk $D(a, r)$ to f. If a holomorphic function is defined over the whole of $\mathbb{C}$ then it is called entire and its Taylor series has an infinite radius of convergence. In particular this means that a holomorphic function is infinitely differentiable on D once it is differentiable one time, a sharp contrast with the theory of real functions.

If a and b are two real numbers then the integral of a function defined over $\mathbb{R}$ is simple: we go from a to b. For a holomorphic function and two points a, $b \in D$ there are *a priori* several different paths γ starting from a and going to b. So, *a priori* the integral

$$\int_{\gamma} f(z) dz$$

depends on the chosen path γ. In fact if γ and γ' are two paths from a to b which can be continuously deformed one into the other then

$$\int_{\gamma} f(z) dz = \int_{\gamma'} f(z) dz.$$

A domain D is called *simply connected* if for any two points z_0 and z_1 and two paths γ, γ' from z_0 to z_1, one can deform γ to γ'. For such domain, the integral $\int_a^b f(z) dz$ is uniquely defined since it is independent on the chosen path from a to b. The domains $\mathbb{C}$, $D(a,r)$, $\mathbb{C} - \mathbb{R}_-$ are simply connected but $\mathbb{C} - \{0\}$ is not simply connected.

One remarkable consequence of the existence of Taylor expansions (2.15) is that a holomorphic function is characterized over the whole of its domain of definition by its values over any disk $D(a,r)$ however small it can be. This implies that if a function f is analytic in a domain $D \subset D'$ with D and D' connected, then one can consider the problem of extending f from D to D'. The problem is that such an extension does not always exist. One such example is the logarithmic function $\ln(z)$. It can be defined over $D = \{x \in \mathbb{C} \ : \ Re\, z > 0\}$ by

$$\ln z = \ln\left(\sqrt{x^2 + y^2}\right) + i \operatorname{Atan}\left(\frac{x}{\sqrt{x^2 + y^2}}\right)$$

and it can be extended to $\mathbb{C} - \mathbb{R}_-$ but not to $\mathbb{C} - \{0\}$. The reason is that in the extension to $\mathbb{C} - \mathbb{R}_-$ one has for $r \in \mathbb{R}_-$

$$\lim_{z \to r} \ln z = \ln(-r) \pm i\pi$$

depending on whether we go to r from the upper half plane $Im\, z > 0$ or from the lower half plane $Im\, z < 0$.

The theory of holomorphic functions can be extended to several complex variables and we can define a holomorphic differential equations to be an equation of the form $f^{(p)}(z) = F(z, f(z), \ldots, f^{(p-1)}(z))$ with F a holomorphic function of several variables. We have a theorem of local existence that is if F is defined over some domain D and $(x_0, y_0^0, y_0^1, \ldots, y_0^{p-1}) \in D$ then there exists a solution f defined in a small neighborhood of D. If one has a path γ between two points z_0 and z_0' then solving the differential equation along this path is actually just like solving a classic real ordinary differential equation.

This implies that for a simply connected domain D, the solution of a differential equation if it exists, that is if no blow-up occurs, is defined uniquely over D. This applies for example to linear differential equations for which the classical theory guarantees the existence of solutions. But for non-simply connected domains, the situation may be more complicated. One basic example is

$$y' = \alpha \frac{1}{z} y$$

with α a real parameter. The equation is defined over $\mathbb{C} - \{0\}$ but it does not admit solution over it if $\alpha \notin \mathbb{Z}$. The reason is that the function that we would like to have as solution, that is

$$y(z) = z^\alpha = \exp\left(\alpha \ln(z)\right)$$

is not defined over $\mathbb{C} - \{0\}$ if $\alpha \notin \mathbb{Z}$.

Chapter 3

Random interval packing

We consider here random sequential packing of interval $[0, 1]$ into the interval $[0, x]$ and we are looking after the limit packing density of this packing. In our presentation, we follow very closely Rényi's method of deriving a differential delay equation, of applying the Laplace transform and then a Tauberian theorem to prove that the limit random packing density is

$$C_R = \int_0^\infty \exp\left\{-2 \int_0^t \frac{1 - e^{-u}}{u} du\right\} dt.$$

Then in Section 3.4 we give another proof of the existence of the limit and the expansion

$$M(x) = C_R x - (1 - C_R) + O\left(\frac{1}{x^n}\right)$$

of the packing density $M(x)$ by using Complex Analysis.

An independent proof of the existence of the limit packing density is available from [Konheim and Flatto (1962)] and we will see in Chapter 5 an existence proof based on integral equations. There are other determination of the limit in the Statistical Mechanics literature [Hemmer (1989); Krapivsky (1992)], which do not use the Laplace transform. After the work of Rényi, many variants have been introduced. Two will be considered in this book: In Chapter 7 a variant of the Kakutani interval splitting algorithm, where the cars may have length lower than 1 but can fit only in intervals of length at least 1. In Chapter 8 we consider the car parking problem with spin. [Ney (1962)] considered the case of a car of varying length and derived integral expressions of the resulting limit packing densities. [Itoh and Mahmoud (2003)] considered the packing where the cars have to park after the last ones.

Among the many other generalizations that exist we present the in time packing result by [Coffman, Jelenković and Poonen (1999); Coffman, Mallows and Poonen (1994); Coffman, Poonen and Winkler (1995); Coffman,

Flatto, Jelenković and Poonen (1998)]. Cars arrive with a Poisson distribution at some place and park if some place is available. If $K(t, x)$ denotes
the expected number of cars parked during $[0, t]$ into $[0, x]$ then it is proved
that

$$\lim_{x \to \infty} \frac{K(t, x)}{x} = \alpha(t) = \int_0^t \exp\left(-2 \int_0^x \frac{1 - e^{-u}}{u} du\right) dx$$

and the variance is also computed in the same paper with more complicated
formulas. Also it is proved that if $N_x(n)$ is the packing density after n
attempts to pack then $N_x(\lceil \lambda x \rceil) \simeq \alpha(\lambda)x$.

3.1   The probabilistic setup of the problem

Let us study the argument by [Rényi (1958)]. Place at random a unit
interval $I_1 = [0, 1]$ into the interval $(0, x)$. The initial point ξ of the interval
I_1 is a random variable uniformly distributed in the interval $(0, x-1)$. Then
place another unit interval at random, independently of the location of the
interval I_1, to the interval $(0, x)$. If the second interval does not intersect
I_1, we keep it and denote by I_2; if it does, we neglect it and choose a new
interval. If the (disjoint) intervals $I_1, I_2, \dots, I_k$ have already been chosen,
the next randomly chosen interval will be kept only if it does not intersect
any of the intervals $I_1, I_2, \dots, I_k$. In this case the interval will be denoted by
I_{k+1}. The procedure is continued as long as possible, i.e. until the following
situation occurs: none of the distances between consecutive intervals, resp.
between the first or last interval and the corresponding endpoint of the
interval $(0, x)$, is greater than 1. See one example of this random packing
process in Figure 3.1. Assume that this occurs after placing N_x intervals.

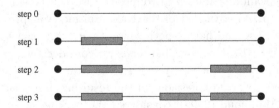

Fig. 3.1   One example of random parking in an interval $[0, x]$

The problem is to determine the expectation $M(x) = E(N_x)$ and the

limit
$$\lim_{x \to \infty} \frac{M(x)}{x} = C_R.$$

Obviously $M(x) = 0$ if $0 \le x \le 1$ and $M(x) = 1$ if $1 < x \le 2$. We will prove in Proposition 3.1 that $M(x)$ is monotone increasing. It is clear that $0 \le M(x) \le x$. We will prove that the above limit exists and determine the value of C_R.

First we show that $M(x)$ satisfies the integral equation

$$M(x+1) = 1 + \frac{2}{x} \int_0^x M(t)dt. \tag{3.1}$$

This can be done as follows: if a unit interval is already placed on the interval $(0, x+1)$ and it is the interval $(t, t+1)$ with $0 \le t \le x$, then the average number of the intervals at the left is $M(t)$ and that of the interval at the right is $M(x - t)$, and thus (since t is uniformly distributed in $(0, x)$)

$$M(x+1) = 1 + \frac{1}{x} \int_0^x M(t) + M(x-t)dt = 1 + \frac{2}{x} \int_0^x M(t)dt. \tag{3.2}$$

Using Equation (3.2), it can be shown that if $x_0 \notin \mathbb{Z}$, then $M(x)$ is infinitely differentiable in a neighborhood of x_0. If $x_0 = n \in \mathbb{Z}$ then f admits $n - 2$ continuous derivative at x_0.

Integrating (3.2), we get

$$M(x) = \begin{cases} 0 & \text{if } x \in [0,1), \\ 1 & \text{if } x \in [1,2), \\ 3 - \frac{2}{x-1} & \text{if } x \in [2,3), \\ 7 - \frac{10}{x-1} - \frac{4}{x-1}\log(x-2) & \text{if } x \in [3,4). \end{cases} \tag{3.3}$$

In [Blaisdell and Solomon (1970)] an expression of $M(x)$ as a series is obtained for $x \in [4, 5)$. Further expressions will most likely be yet harder to obtain.

Those few expressions make it clear that one cannot expect to be able to derive the limit of $\frac{M(x)}{x}$, if it exists, by such explicit computations. Something else is needed that avoids the problem of the intervals $[n, n+1)$, it will turn out that the Laplace transform allows us to derive the solution of the problem.

Multiplying Equation (3.1) by x and differentiating both sides with respect to x we get that $M(x)$ satisfies the "delay differential equation"

$$xM'(x+1) + M(x+1) = 2M(x) + 1 \text{ with } x > 0. \tag{3.4}$$

This equation can be used to prove directly that M is increasing:

Proposition 3.1. *[Rényi (1958)] The function $M(x)$ is monotone increasing on $\mathbb{R}_+$.*

Proof. By a change of variable, Equation (3.4) is rewritten as

$$M(x+1) = 1 + 2 \int_0^1 M(tx) dt.$$

As a consequence we deduce that if M is increasing on $[0, x_0]$ for any $x_0 > 0$ then it is increasing on $[0, x_0 + 1]$ as well. Starting from $x_0 = 1$ the result follows by induction. $\square$

3.2 The solution of the delay differential equation using Laplace transform

Let

$$\varphi(s) = \int_0^\infty M(x) e^{-sx} dx \qquad (3.5)$$

be the Laplace transform of the packing function M. The existence of the Laplace transform (3.5) for $s = \sigma + it$, $\sigma > 0$ follows from the inequality $0 \leq M(x) \leq x$.

Theorem 3.1. *[Rényi (1958)] The function $\varphi(s)$ has the expression*

$$\varphi(s) = \frac{e^{-s}}{s^2} \int_s^\infty \exp\left\{ -2 \int_s^t \frac{1 - e^{-u}}{u} du \right\} dt. \qquad (3.6)$$

Proof. The function $M(x)$ satisfies the delayed differential equation

$$xM'(x+1) + M(x+1) = 2M(x) + 1 \text{ for } x > 0, \qquad (3.7)$$

and the initial condition

$$M(x) = 0 \text{ for } 0 \leq x < 1.$$

Multiplying both sides of (3.7) by e^{-sx} and integrating we get

$$\int_0^\infty xM'(x+1) e^{-sx} dx + \int_0^\infty M(x+1) e^{-sx} dx = 2\varphi(s) + \frac{1}{s}.$$

Since $M(x) = 0$ if $0 \leq x \leq 1$, we have

$$\int_0^\infty M(x+1) e^{-sx} dx = e^s \int_1^\infty M(t) e^{-st} dt = e^s \varphi(s).$$

Let us now consider the integral of the differential. The function M is constant equal to 1 on $[1, 2)$, so if one restricts M to the interval $[1, \infty)$, then M is differentiable at 1 of value 0. We then have by integration by part

$$\int_0^\infty x M'(x+1) e^{-sx} dx = -\frac{d}{ds} \left(\int_0^\infty M'(x+1) e^{-sx} dx \right)$$
$$= -\frac{d}{ds} \left([M(x+1) e^{-sx}]_0^\infty \right.$$
$$\left. - \int_0^\infty M(x+1)(-s) e^{-sx} dx \right)$$
$$= -\frac{d}{ds} \left(-M(1) + s \int_0^\infty M(x+1) e^{-sx} dx \right)$$
$$= -\frac{d}{ds} \left(-M(1) + s e^s \varphi(s) \right) = -\frac{d}{ds} (s e^s \varphi(s)).$$

Combining the preceding relations we get that $\varphi(s)$ satisfies the differential equation

$$\frac{d}{ds} (s e^s \varphi(s)) = \varphi(s)(e^s - 2) - \frac{1}{s}.$$

Introducing the notation $w(s) = e^s \varphi(s)$, we get that $w(s)$ satisfies the differential equation

$$s w'(s) = -2 w(s) e^{-s} - \frac{1}{s}. \tag{3.8}$$

The estimate $0 \le M(x) < x$ implies for positive s the estimation

$$0 \le \varphi(s) = \int_1^\infty M(x) e^{-sx} dx \le \int_1^\infty x e^{-sx} dx = e^{-s} \left(\frac{1}{s} + \frac{1}{s^2} \right)$$

and thus

$$\lim_{s \to \infty} w(s) = 0. \tag{3.9}$$

Equation (3.8) is a first order linear inhomogeneous differential equation, for which we use the variation of constant method of Chapter 2. First the vector space of solutions of the homogeneous equation

$$s w'(s) = -2 w(s) e^{-s}$$

is generated by

$$w_0(s) = \exp \left(-2 \int_\infty^s \frac{e^{-u}}{u} du \right).$$

Then, one write the solution $w(s)$ in the inhomogeneous Equation (3.8) as $\alpha(s) w_0(s)$ and gets by insertion in Equation (3.8)

$$\alpha'(s) = -\frac{1}{s^2} \exp \left(2 \int_\infty^s \frac{e^{-u}}{u} du \right).$$

Clearly $\alpha'(s)$ is integrable on $(0, \infty)$ and so one gets for some constant α_0:

$$w(s) = \left\{ \alpha_0 + \int_\infty^s -\frac{1}{t^2} \exp \left(2 \int_\infty^t \frac{e^{-u}}{u} du \right) dt \right\} w_0(s)$$
$$= \alpha_0 w_0(s) + \int_\infty^s \frac{1}{t^2} \exp \left(-2 \int_\infty^s \frac{e^{-u}}{u} du + 2 \int_\infty^t \frac{e^{-u}}{u} du \right) dt$$
$$= \alpha_0 w_0(s) + \frac{1}{s^2} \int_s^\infty \exp \left(-2 \int_s^t \frac{1 - e^{-u}}{u} du \right) dt.$$

The limit (3.9) implies $\alpha_0 = 0$ and thus the expression (3.6).   □

There exist inversion formulas for getting back a function from its Laplace transform. But we should refrain from using them directly and trying to compute $M(x)$ exactly: we would only get the values of Equation (3.3) and not the limit that we are looking after. Instead, we will use a Tauberian theorem for computing the limit of $\frac{M(x)}{x}$.

3.3   The computation of the limit

In this section, we prove that the limit $\lim_{x \to \infty} \frac{M(x)}{x}$ exists and we compute its value. The proof uses what is called "Tauberian theorems". Such theorems were devised by Hardy-Littlewood and Tauber and later developed into an extensive theory by [Wiener (1932)]. A recent monograph is [Korevaar (2004)]. Tauberian theorems are generally not proved in undergraduate courses because they are too hard and not in graduate course since analytical methods are generally more powerful. So, we provide a proof adapted from [Hardy-Littlewood (1930)].

Theorem 3.2. *If f is positive and integrable over every finite interval $(0, T)$ and $e^{-st} f(t)$ is integrable over $(0, \infty)$ for any $s > 0$ and if*

$$g(s) = \int_0^\infty e^{-st} f(t) dt \simeq H s^{-\beta} \quad \text{as } s \to 0$$

where $\beta > 0$ and $H > 0$ then when $x \to \infty$, we have

$$F(x) = \int_0^x f(t) dt \simeq \frac{H}{\Gamma(1 + \beta)} x^\beta \quad \text{as } x \to \infty.$$

Proof. By integrating by part we get

$$g(s) = s \int_0^\infty e^{-st} F(t) dt.$$

Let us write $h(s) = \frac{g(s)}{s}$. Now the function $h(s)$ is infinitely differentiable and one gets

$$(-1)^n h^{(n)}(s) = \int_0^\infty t^n e^{-st} F(t) dt.$$

So, the sign of $h^{(n)}(s)$ is $(-1)^n$. We now use the following Landau's lemma (See [Hardy-Littlewood (1914)], page 175):

Lemma 3.1. *If f is a real differentiable increasing function from $\mathbb{R}_+$ with*

$$f(x) \simeq x^a \text{ for some } a > 0 \text{ when } x \to \infty$$

then

$$f'(x) \simeq ax^{a-1} \text{ when } x \to \infty.$$

As a consequence we get for all n as $s \to 0$:

$$(-1)^n h^{(n)}(s) \simeq H(\beta+1)\ldots(\beta+n)s^{-\beta-1-n}.$$

As n goes to ∞, we have the well-known equivalent

$$\Pi_{i=1}^n (\beta+i) \simeq n! \frac{n^\beta}{\beta} \frac{1}{\Gamma(\beta)}$$
$$\simeq n^n e^{-n} \sqrt{2\pi n} \frac{n^\beta}{\beta} \frac{1}{\Gamma(\beta)} \tag{3.10}$$

with Γ the Gamma function.

Let us write $\phi(w) = we^{-w}$ and get

$$(-1)^n h^{(n)}(s) = \left(\frac{n}{s}\right)^{n+1} \int_0^\infty \phi(w)^n F\left(\frac{n}{s}w\right) dw. \tag{3.11}$$

For all n the expression of the derivative in (3.11) depends on $\phi(w)^n$. The function $\phi(w)$ attains its maximum at 1 and thus most of the integral is concentrated around it. So, we will be able to prove that $(-1)^n h^{(n)}(s)$ is comparable to $F(\frac{n}{s})$ and get the equivalent of F.

Let us fix an $\epsilon \in (0, \frac{1}{2}]$ and define the following integrals:

$$I_{n,0} = \int_0^{\frac{n}{s}(1-\epsilon)} t^n e^{-st} dt, \quad I_{n,1} = \int_{\frac{n}{s}(1-\epsilon)}^{\frac{n}{s}(1+\epsilon)} t^n e^{-st} dt,$$
$$\text{and} \quad I_{n,2} = \int_{\frac{n}{s}(1+\epsilon)}^\infty t^n e^{-st} dt.$$

We also write

$$J_{n,0} = \int_0^{\frac{n}{s}(1-\epsilon)} t^n e^{-st} F(t) dt, \quad J_{n,1} = \int_{\frac{n}{s}(1-\epsilon)}^{\frac{n}{s}(1+\epsilon)} t^n e^{-st} F(t) dt,$$
$$\text{and} \quad J_{n,2} = \int_{\frac{n}{s}(1+\epsilon)}^\infty t^n e^{-st} F(t) dt.$$

We have the equalities

$$I_{n,0} + I_{n,1} + I_{n,2} = \int_0^\infty t^n e^{-st} dt = \frac{n!}{s^{n+1}}$$

and $J_{n,0} + J_{n,1} + J_{n,2} = (-1)^n h^{(n)}(s)$.

We can write $I_{n,0} = \left(\frac{n}{s}\right)^{n+1} \int_0^{1-\epsilon} \phi(w)^n dw$ and similarly for $I_{n,1}$ and $I_{n,2}$. Since $\phi(w)$ attains its maximum at 1, one can prove that there exists n_0 such that $n \geq n_0$ implies

$$I_{n,0} \leq \epsilon I_{n,1} \text{ and } I_{n,2} \leq \epsilon I_{n,1}.$$

This inequality implies by the monotonicity of F the inequality

$$J_{n,0} \leq I_{n,0} F\left(\frac{n}{s}(1-\epsilon)\right)$$
$$\leq \epsilon I_{n,1} F\left(\frac{n}{s}(1-\epsilon)\right)$$
$$\leq \epsilon J_{n,1}.$$

Since $\phi(w) < e^{-1}$ for $w \geq 1 + \epsilon$, there exists a constant $C < e^{-1}$ and $\alpha > 0$ such that

$$\phi(w) \leq Ce^{-\alpha w} \text{ for } w \geq 1 + \epsilon.$$

Henceforth we get

$$
\begin{aligned}
J_{n,2} &= \int_{\frac{n}{s}(1+\epsilon)}^{\infty} t^n e^{-st} F(t) dt \\
&= \left(\frac{n}{s}\right)^n \int_{\frac{n}{s}(1+\epsilon)}^{\infty} \phi\left(\frac{s}{n}t\right)^n F(t) dt \\
&\leq \left(\frac{n}{s}\right)^n \int_{\frac{n}{s}(1+\epsilon)}^{\infty} C^n e^{-\alpha st} F(t) dt \\
&\leq \left(\frac{n}{s}\right)^n C^n \int_0^{\infty} e^{-\alpha st} F(t) dt = \left(\frac{n}{s}\right)^n C^n h(\alpha s).
\end{aligned}
$$

So, using the equivalence of g we get for say $s \leq s_0$

$$
\begin{aligned}
J_{n,2} &\leq \left(\frac{n}{s}\right)^n C^n \frac{g(\alpha s)}{\alpha s} \\
&\leq \left(\frac{n}{s}\right)^n C^n 2 \frac{H}{\alpha} s^{-\beta-1} \\
&\leq s^{-\beta-1-n} n^n C^n \frac{2H}{\alpha}.
\end{aligned}
$$

Since $C < e^{-1}$, the term $n^n C^n$ is dominated by the term $n! n^\beta$ of the equivalent (3.10). So, there exists $n_1 \geq n_0$ such that for $n \geq n_1$ there is $s_1(n)$ with for $s \leq s_1(n)$:

$$(1 - 2\epsilon) J_{n,1} \leq H n^n e^{-n} \sqrt{2\pi n} n^\beta \frac{1}{\Gamma(\beta+1)} s^{-\beta-1-n} \leq (1 + 2\epsilon) J_{n,1}.$$

On the other hand we have by monotonicity of F the estimation

$$F\left(\frac{n}{s}(1-\epsilon)\right) I_{n,1} \leq J_{n,1} \leq F\left(\frac{n}{s}(1+\epsilon)\right) I_{n,1}.$$

The integral $I_{n,1}$ is estimated by $n^n e^{-n} \sqrt{2\pi n} s^{-n-1}$. Putting it altogether this means that for all $\epsilon > 0$, there exist n_2 and $s_2(n)$ such that for $s \leq s_2(n)$.

$$(1 - \epsilon) F\left(\frac{n}{s}\right) \leq H \frac{1}{\Gamma(\beta+1)} \frac{n^\beta}{s^\beta} \leq (1 + \epsilon) F\left(\frac{n}{s}\right).$$

Fixing $n = n_2$ and writing $x = \frac{n}{s}$ we get the required equivalent. □

Theorem 3.3. *[Rényi (1958)] One has the limit*

$$\lim_{x \to \infty} \frac{M(x)}{x} = C_R \qquad (3.12)$$

with

$$C_R = \int_0^{\infty} \exp\left\{-2 \int_0^t \frac{1 - e^{-u}}{u} du\right\} dt. \qquad (3.13)$$

Proof. Formula (3.6) implies that

$$\lim_{s \to +0} s^2 \varphi(s) = C_R. \tag{3.14}$$

If one applies Theorem 3.2 to $M(x)$, one gets using Equation (3.14) and (3.13) the limit

$$\lim_{x \to \infty} \frac{1}{x^2} \int_0^x M(t) dt = \frac{C_R}{2}. \tag{3.15}$$

Now divide both sides of (3.1) by $x + 1$ and let x tend to infinity. Equation (3.15) implies the limit (3.12). □

In [Rényi (1958)] it is remarked that one can get other expressions of C_R. By integration by part one gets:

$$
\begin{aligned}
C_R &= 2 \int_0^\infty (1 - e^{-t}) \exp \left\{ -2 \int_0^t \frac{1 - e^{-u}}{u} du \right\} dt \\
&= 2C_R - 2 \int_0^\infty \exp \left\{ -t - 2 \int_0^t \frac{1 - e^{-u}}{u} du \right\} dt.
\end{aligned}
\tag{3.16}
$$

Thus

$$C_R = 2 \int_0^\infty \exp \left\{ -t - 2 \int_0^t \frac{1 - e^{-u}}{u} du \right\} dt. \tag{3.17}$$

This kind of formula will reappear in Chapters 7 and 8. This formula is much more efficient for the effective computation of C_R because the integral converges faster but is still slow compared to methods based on Dvoretzky-Robbins theory [Blaisdell and Solomon (1970)] exposed in Chapter 5. The result of those computations is that $C_R = 0.7475979202\ldots$.

3.4 Packing on circle and the speed of convergence

In Theorem 3.4 below we give an expansion of the packing density $M(x)$ in the form

$$M(x) = C_R x - (1 - C_R) + O \left(\frac{1}{x^n} \right).$$

In Chapter 5 we will improve this estimation of the error term. The rapidity of the convergence to the affine function is apparent in Figure 3.2 and has been used in [Blaisdell and Solomon (1970)] to compute the limit packing density C_R efficiently, that is without using the integral expressions (3.13) and (3.17).

What is interesting is that one can rewrite the above equation as

$$\frac{M(x)+1}{x+1} = C_R + O\left(\frac{1}{x^n}\right).$$

The left hand side is the density of a packing of intervals $[0,1]$ in a circle of length $x + 1$: after the first interval is put we are reduced to a packing on the interval of length x; this was remarked for example in [Blaisdell and Solomon (1970)].

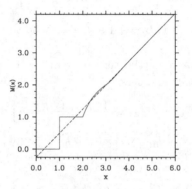

Fig. 3.2   The expectation $M(x)$ and the affine fitting $C_R x - (1 - C_R)$

Theorem 3.4. *[Rényi (1958)] For all $n \in \mathbb{N}$ we have*

$$M(x) = C_R x - (1 - C_R) + O\left(\frac{1}{x^n}\right)$$

as x goes to infinity.

Proof.   Let us first determine the limit of $M(x) - C_R x$:

$$\lim_{x \to \infty} M(x) - C_R x = -(1 - C_R).$$

To this aim we make use of the fact that $\varphi(s)$, the Laplace transform of $M(x)$ can be written by (3.6) as follows:

$$\varphi(s) = \frac{\Phi(s)}{s^2}.$$

Here $\Phi(s)$ can be written in the following form

$$\Phi(s) = e^{-s}\left(C_R \exp\left\{2\int_0^s \frac{1-e^{-u}}{u}\,du\right\} - \int_0^s \exp\left\{2\int_t^s \frac{1-e^{-u}}{u}\,du\right\} dt\right).$$

Since $(1 - e^{-z})/z$ is an entire function, $\Phi(s)$ is easily seen to be also an entire function. This implies that

$$\varphi(s) = \frac{C_R}{s^2} + \frac{\Phi'(0)}{s} + \psi(s),$$

where

$$\psi(s) = \frac{\Phi(s) - C_R - \Phi'(0)s}{s^2}$$

is also an entire function. Now a simple calculation shows that

$$\Phi'(0) = C_R - 1$$

thus

$$\varphi(s) = \frac{C_R}{s^2} - \frac{1 - C_R}{s} + \psi(s)$$

where $\psi(s)$ is an entire function.

Now according to the inversion formula (3.23) for Laplace transform given in Appendix 3.5 we have

$$M(x) = \frac{1}{2\pi i} \int_{\sigma - i\infty}^{\sigma + i\infty} e^{xs} \varphi(s) ds \quad \text{for} \quad \sigma > 0$$

and thus

$$M(x) = C_R x - (1 - C_R) + \frac{1}{2\pi i} \int_{\sigma - i\infty}^{\sigma + i\infty} e^{xs} \psi(s) ds \quad \text{for} \quad \sigma > 0. \tag{3.18}$$

Since $\psi(s)$ is an entire function, in (3.18) one can also take $\sigma = 0$ and obtain by writing $s = it$ that

$$M(x) = C_R x - (1 - C_R) + \frac{1}{2\pi} \int_{-\infty}^{\infty} e^{ixt} \psi(it) dt. \tag{3.19}$$

For $s = it$ (t is real) in the integral

$$\int_{s}^{\infty} \exp\left\{ -2 \int_{0}^{t} \frac{1 - e^{-u}}{u} du \right\} dt$$

occurring in the formula

$$\varphi(s) = \frac{e^{-s}}{s^2} \exp\left\{ 2 \int_{0}^{s} \frac{1 - e^{-u}}{u} du \right\} \int_{s}^{\infty} \exp\left\{ -2 \int_{0}^{t} \frac{1 - e^{-u}}{u} du \right\} dt$$

one can choose the imaginary axis as the path of integration. We have seen in Appendix 2.6 that one can deform the path of integration while keeping the same endings. Thus to change the endings at infinity, it is enough to show, that writing K_R for the quarter circle $z = Re^{i\phi}$, $0 \le \phi \le \pi/2$, one has

$$\lim_{R \to \infty} A(R) = 0 \text{ with } A(R) = \int_{K_R} \exp\left\{-2 \int_0^z \frac{1 - e^{-u}}{u} du\right\} ds. \quad (3.20)$$

In order to prove (3.20) we first estimate $A(R)$. We have

$$|A(R)| = \left| R \int_0^{\pi/2} \exp\left\{-2 \int_0^R \frac{1 - e^{-re^{i\phi}}}{r} dr\right\} d\phi \right|$$
$$\le R \int_0^{\pi/2} \exp Re \left\{-2 \int_0^R \frac{1 - e^{-re^{i\phi}}}{r} dr\right\} d\phi$$
$$\le R \int_0^{\pi/2} \exp\left\{-2 \int_0^R \frac{1 - e^{-r\cos\phi}\cos(r\sin\phi)}{r} dr\right\} d\phi$$
$$\le R \int_0^{\pi/2} \exp\left\{-2 \int_1^R \frac{1 - e^{-r\cos\phi}\cos(r\sin\phi)}{r} dr\right\} d\phi$$
$$\le \frac{1}{R} \int_0^{\pi/2} \exp\left\{2 \int_1^R \frac{e^{-r\cos\phi}\cos(r\sin\phi)}{r} dr\right\} d\phi,$$

where we have used the formulas $|e^z| = e^{Re\, z}$ and

$$\int_0^{Re^{i\phi}} \frac{f(z)}{z} dz = \int_0^R \frac{f(re^{i\phi})}{r} dr.$$

Now, if $0 \le \phi \le \frac{\pi}{4}$, then

$$\int_1^R \frac{e^{-r\cos\phi}\cos(r\sin\phi)}{r} dr \le \int_1^\infty e^{-\frac{r}{\sqrt{2}}} \le 3$$

and if $\frac{\pi}{4} \le \phi \le \frac{\pi}{2}$, then

$$\int_1^R \frac{e^{-r\cos\phi}\cos(r\sin\phi)}{r} dr \le \int_1^{\frac{\pi}{2\sin\phi}} \frac{e^{-r\cos\phi}\cos(r\sin\phi)}{r} dr \le \frac{\pi}{2\sin\phi} \le 3,$$

because $\frac{1}{r} e^{-r\cos\phi}$ is a monotone decreasing function. Thus,

$$\left| \int_{K_R} \exp\left\{-2 \int_0^z \frac{1 - e^{-u}}{u} du\right\} dz \right| \le \frac{e^6 \pi}{2R}$$

and so (3.20) holds and hence we have

$$\int_{it}^\infty \exp\left\{-2 \int_0^u \frac{1 - e^{-v}}{v} dv\right\} du = i \int_t^\infty \exp\left\{-2 \int_0^u \frac{1 - e^{-iv}}{v} dv\right\} du.$$

Thus, if t is positive, then $\varphi(it)$ can be written as

$$\varphi(it) = -i \frac{e^{-it}}{t^2} \exp\left\{2 \int_0^t \frac{1 - e^{-iu}}{u} du\right\} \int_t^\infty \exp\left\{-2 \int_0^u \frac{1 - e^{-iv}}{v} dv\right\} du. \quad (3.21)$$

Obviously $|\varphi(-it)| = |\varphi(it)|$, thus investigating the asymptotic behavior of $\varphi(it)$ and $\psi(it)$ we can restrict ourselves to the positive values of t. An integration by parts shows that the integral

$$\int_1^\infty \frac{e^{-iu}}{u} du$$

exists.

One gets the estimations

$$\exp\left\{2 \int_0^t \frac{1-e^{-iu}}{u} du\right\} = O\left(|t|^2\right)$$

and

$$\int_t^\infty \exp\left\{-2 \int_0^u \frac{1-e^{-iv}}{v} dv\right\} du = O\left(\frac{1}{|t|}\right)$$

as $|t| \to \infty$. Those estimates imply

$$\varphi(it) = O\left(\frac{1}{|t|}\right) \text{ and } \psi(it) = O\left(\frac{1}{|t|}\right)$$

as $t \to \infty$.

Thus it follows from (3.19) using integration by parts that

$$M(x) - C_R x + (1-C_R) = -\frac{1}{2\pi i(x-1)} \int_{-\infty}^\infty e^{it(x-1)} \frac{d}{dt}\left(e^{it}\psi(it)\right) dt.$$

Since

$$\psi(it) = \varphi(it) + \frac{C_R}{t^2} - \frac{i(1-C_R)}{t},$$

we have

$$\frac{d}{dt}\left(e^{it}\psi(it)\right) - \frac{d}{dt}\left(e^{it}\varphi(it)\right) + \frac{(1-C_R)e^{it}}{t} + O\left(\frac{1}{t^2}\right).$$

Equation (3.21) (or Equality (3.8)) shows that

$$\frac{d}{dt}(e^{it}\varphi(it)) = O\left(\frac{1}{t^2}\right).$$

Using also the estimations

$$\left|\int_1^\infty \frac{e^{ixt}}{t} dt\right| \le K$$

and

$$\left|\int_{-\infty}^{-1} \frac{e^{ixt}}{t} dt\right| \le K,$$

which are valid if $x \geq 1$ with K a constant independent of x, one gets

$$M(x) - C_R x + (1 - C_R) = O\left(\frac{1}{x}\right).$$

By induction one can deduce from the differential equation (3.8) the estimation

$$\frac{d^n}{dt^n}\left(e^{it}\varphi(it)\right) = O\left(\frac{1}{t^2}\right),$$

thus repeating the above argument we get by an integration by part the expression

$$M(x) - C_R x + (1 - C_R) = -\frac{1}{2\pi i^n (x-1)^n}\int_{-\infty}^{\infty} e^{it(x-1)}\frac{d^n}{dt^n}\left(e^{it}\psi(it)\right) dt$$

from which the general result follows.   □

3.5   Appendix: The Laplace transform

The Laplace transform is a widely used integral transform in electrical engineering, signal processing, physics and in our case probability theory. One of its main advantages is that it transforms derivation and integration of a function f into multiplication and division, which makes it very useful for solving differential equations. In contrast to the Fourier transform to which it is related the Laplace transform is applied to integrable functions in order to study transient phenomena.

Here we denote by $L^1(\mathbb{R}_+)$ the space of absolutely integrable functions over $\mathbb{R}_+$. If $f \in L^1(\mathbb{R}_+)$, then the Laplace transform is defined as

$$F(s) = \mathcal{L}\{f(t)\} = \int_0^{\infty} f(t)e^{-st}dt. \tag{3.22}$$

The parameter s is generally a complex number $s = x + iy$.

The function f is chosen integrable to ensure that the integral defining $F(s)$ converge for $Re\, s \geq 0$. We will encounter some examples that do not satisfy this condition but converge for $Re\, s > s_0$ for some $s_0 \in \mathbb{R}$. This does not change the essence of what is stated there. The Laplace transform is defined over this half-space and it is easy to prove that it is holomorphic over it.

The main properties of the Laplace transform are listed below. It is linear in the function f, i.e.

$$\mathcal{L}\{af + bg\} = a\mathcal{L}\{f\} + b\mathcal{L}\{g\}.$$

By differentiating Equation (3.22) with respect to s under the integral sign, we get

$$F'(s) = \int_0^\infty (-t) f(t) e^{-st} dt.$$

Thus we get $\mathcal{L}\{tf(t)\} = -\mathcal{L}\{f(t)\}'$. Assuming that f and f' are absolutely integrable, we get by integration by parts the formula

$$\mathcal{L}\{f'\}(s) = s\mathcal{L}\{f\}(s) - f(0).$$

Another property, which is well suited to the integration of differential delay equations is that if $f = 0$ on $[0, a]$ then

$$\mathcal{L}\{f(t)\} = e^{-as} \mathcal{L}\{f(a+t)\}.$$

In order to state the next property, we need to define another integral transformation:

Definition 3.1. If $f, g \in L^1(\mathbb{R}_+)$ then we define their convolution $f \star g$ by

$$(f \star g)(t) = \int_0^t f(y) g(t-y) dy.$$

One can prove that we have $f \star g \in L^1(\mathbb{R}_+)$.

By using the convolution product, we can get an algebra product on $L^1(\mathbb{R}_+)$. But it is an algebra with no unit element since for an unit e we would have for all $x \geq 0$ and all integrable functions the equality

$$f(x) = \int_0^x e(t) f(x-t) dt.$$

Thus e would have to be concentrated on 0, which is impossible. Sometimes one way to deal with that is to use the Dirac δ_0, which is a distribution (i.e. not a function) and satisfies the above equation. But fortunately we will not need this δ_0 a lot. By a change of variables in the integral we get

$$\mathcal{L}\{(f \star g)(t)\} = \mathcal{L}\{f(t)\} \mathcal{L}\{g(t)\}.$$

So, the convolution product is transformed into ordinary product.

Remark 3.1. If for the sake of an example we allow the measure $f(x)dx$ to be replaced by a measure $\mu(dx)$ with $\mu(\{n\}) = p_n$ for some values p_n then we get

$$\begin{aligned}
\mathcal{L}(\mu)(s) &= \int_0^\infty e^{-sx} \mu(dx) \\
&= \sum_{n=0}^\infty e^{-sn} p_n \\
&= \sum_{n=0}^\infty \alpha^n p_n
\end{aligned}$$

with $\alpha = e^{-s}$. This explains the close analogy between the properties of Laplace transforms and generating functions. We will not need anymore Laplace transform for measures.

Theorem 3.5. *If* $f, g \in L^1(\mathbb{R}_+)$ *then* $\mathcal{L}(f) = \mathcal{L}(g)$ *implies* $f = g$.

Proof.   There are many ways to prove this result. One of them is to write $f_2(t) = F(it)$, $g_2(t) = G(it)$ and remark that f_2, g_2 are Fourier transforms of f and g. $\square$

Actually, one can prove more, that is that F is defined by its values on an interval $[a, \infty)$ for any $a > 0$.

More problematic is to obtain back the function f from the Laplace transform. The inversion formula for Laplace transform is known as Mellin inverse formula and is

$$f(t) = \mathcal{L}^{-1}\{F\}(t) = \frac{1}{2\pi i} \int_{\gamma - i\infty}^{\gamma + i\infty} e^{st} F(s) ds \qquad (3.23)$$

where γ is a real number so that the contour path of integration is in the region where the Laplace transform is defined. If f is integrable then γ can be chosen to be 0.

Also, we have two limit relations:

$$f(0^+) = \lim_{s \to \infty} sF(s) \text{ and } \lim_{x \to \infty} f(x) = \lim_{s \to 0} sF(s)$$

which are useful in asymptotic analysis.

Chapter 4

On the minimum of gaps generated by 1-dimensional random packing

Let $L(t)$ be the random variable which represents the minimum of length of gaps generated by random packing of unit intervals into $[0, t]$. We study the probability that $L(x) \geq h$ as x goes to infinity by giving an integral equation describing it.

By using Renewal Theory and Laplace Transform methods it is proved in Sections 4.1 and 4.2 that the probability $f(x) = Pr(L(x) \geq h)$ is well approximated by an exponential function (see Figure 4.1).

The exponential decay $a(h)$ is expressed as an explosion time of an ordinary differential equation, which is obtained from the Laplace transform in Section 4.2. Rigorous numerical calculations of $a(h)$ are performed in Section 4.3 by analyzing this singular differential equation and proving some lower and upper bounds for its solution. In Section 4.4 we consider the asymptotic behavior of $a(h)$. We first prove some basic estimates by using Renewal Theory in the case $h \to 1$ and then, following [Morrison (1987)], we consider the more powerful Laplace transform methods which allow to find the complete expansion in the case $h \to 1$ and the first order in the case $h \to 0$. In Section 4.6 we summarize the main facts about Renewal Theory that we need. The numerical solution to this problem was suggested by J. M. Hammersley.

4.1 Main properties of $Pr(L(x) \geq h)$

Let $L(x)$ be the random variable which represents the minimum of lengths of gaps generated by the above random packing procedure. If a unit interval is already placed on the interval $[0, x + 1]$ with initial point at y, then by

the independence
$$Pr(L(x+1) \geq h|y) = Pr(L(y) \geq h)Pr(L(x-y) \geq h).$$
Since y is uniformly distributed in $[0, x]$, we have
$$Pr(L(x+1) \geq h) = \frac{1}{x} \int_0^x Pr(L(y) \geq h)Pr(L(x-y) \geq h)dy \qquad (4.1)$$
with
$$Pr(L(x) \geq h) = \begin{cases} 0 \text{ for } 0 \leq x < h, \\ 1 \text{ for } h \leq x < 1, \\ 0 \text{ for } x = 1. \end{cases}$$

Put
$$f(x) = Pr(L(x) \geq h) \qquad (4.2)$$
then $f(x)$ satisfies
$$f(x+1) = \frac{1}{x} \int_0^x f(x-y)f(y)dy \qquad (4.3)$$
with
$$f(x) = \begin{cases} 0 \text{ for } 0 \leq x < h, \\ 1 \text{ for } h \leq x < 1, \\ 0 \text{ for } x = 1. \end{cases} \qquad (4.4)$$
We define the functions
$$f_k(x) \qquad \text{for} \qquad k = 1, 2, 3, \cdots,$$
as follows;
$$f_1(x) = \begin{cases} 1 \text{ for } h \leq x < 1, \\ 0 \text{ for } x < h \text{ or } 1 \leq x, \end{cases}$$
and for $k \geq 2$,
$$f_k(x) = \begin{cases} \dfrac{1}{x-1} \displaystyle\sum_{i+j=k} f_i \star f_j(x-1) \text{ for } x \neq 1, \\ 0 \qquad\qquad\qquad\qquad\qquad \text{for } x = 1, \end{cases}$$
where $\star$ is the convolution product defined in Section 3.5.

We define the probability densities $P_h^{k\star}(x)$ as the k-fold convolution of $P_h(x)$, where
$$P_h(x) = \begin{cases} \dfrac{1}{1-h} & \text{if } h \leq x \leq 1, \\ 0 & \text{otherwise.} \end{cases}$$

Lemma 4.1. *One has:*
$$Pr(L(x) \geq h) = \sum_{i=1}^{\infty} f_i(x).$$

Proof.   The existence and uniqueness of the solution of Equation (4.1) is trivial, since the functional equation gives a unique way of constructing the solution from the values of $f(x)$ for $x \in [0,1]$.

$$f(x+1) = \frac{1}{x}\left(\sum_{i=1}^{\infty} f_i\right) \star \left(\sum_{j=1}^{\infty} f_j\right)(x) = f_2(x+1) + f_3(x+1) + \cdots.$$

Since $f_1(x+1) = 0$ for $x > 0$, $f(x) = \sum_{i=1}^{\infty} f_i(x)$ satisfies Equation (4.1). We have $f_k(x) > 0$ for $k(1+h) - 1 < x < 2k - 1$, and otherwise $f_k(x) = 0$.
   So, if $k(1+h) - 1 < x < (k+1)(1+h) - 1$, $f(x) = \sum_{i=1}^{k} f_i(x)$. Hence

$$f(x) = \sum_{i=1}^{\lceil (x+1)/(1+h)\rceil} f_i(x)$$

is the solution of Equation (4.1).   □

Lemma 4.2. *One has:*

$$\frac{(1-h)^k}{2^{k-1}} P_h^{k\star}(x - k + 1) \le f_k(x).$$

Proof.   The statement for $k = 1$ follows from the definition. Assume that the statement holds for every positive integer k for which $k < k_0$. Then

$$f_{k_0+1}(x) = \frac{1}{x-1}\int_0^{x-1} \sum_{i+j=k_0+1} f_i(x-1-y)f_j(y)dy$$

$$= \frac{1}{x-1}k_0 \frac{(1-h)^{k_0+1}}{2^{k_0-1}} P_h^{(k_0+1)\star}(x - k_0).$$

Since $P_h^{(k_0+1)\star}(x - k_0) = 0$ for $2k_0 + 1 \le x$, we have

$$f_{k_0+1}(x) \ge \frac{1}{2k_0}k_0 \frac{(1-h)^{k_0+1}}{2^{k_0-1}} P_h^{(k_0+1)\star}(x - k_0)$$
$$\ge \frac{(1-h)^{k_0+1}}{2^{k_0}} P_h^{(k_0+1)\star}(x - k_0),$$

which proves the result by induction.   □

Lemma 4.3. *If $0 < h < 1$ then there is a constant $c > 0$ such that*

$$1 > \lim_{x\to\infty} \frac{e^{-cx}}{f(x)},$$

i.e. we have a lower estimate $Pr(L(x) \ge h) \ge Ce^{-cx}$ for some $C > 0$ and x sufficiently large.

Proof.   Put

$$Q_h(x) = \frac{1-h}{2} P_h(x-1) \text{ and } u(x) = (1-h)P_h(x).$$

We have

$$(1-h)P_h \star Q_h^{(k-1)\star}(x) \leq f_k(x).$$

We define the function $Z(x)$ by

$$Z(x) = \sum_{i=1}^{\infty} (1-h)P_h \star Q_h^{(i-1)\star}(x)$$

and we have the relation

$$Z(x) \leq \sum_{i=1}^{\infty} f_i(x) = f(x).$$

But obviously we have the relation

$$
\begin{aligned}
u(x) + Z \star Q_h(x) &= u(x) + \sum_{i=1}^{\infty} u \star Q_h^{i\star}(x) \\
&= \sum_{i=0}^{\infty} u \star Q_h^{i\star}(x) = Z(x),
\end{aligned}
$$

where we have used the convention $Q_h^{0\star}(x) = 1$. From this we conclude that $Z(x)$ is solution of the renewal equation

$$u + Z \star Q_h = Z.$$

By using Theorem 4.11 from the theory of renewal equations, we get $Z(x) \simeq ae^{-kx}$ for some constants $a, k > 0$ (see Section 4.6). Thus we get an exponential lower bound on $Pr(L(x) \geq h)$. $\square$

Note that in order to prove the existence of the exponential decay, another strategy is to use directly the differential equation given in the next section and prove the existence of singularity for its solutions.

4.2   Laplace transform of $Pr(L(x) \geq h)$

Denote by $g(s)$ the Laplace transform of $f(x+1)$:

$$g(s) = \int_0^{\infty} e^{-sx} f(x+1) dx.$$

Theorem 4.1. *If* $0 < h < 1$ *then:*

   (i) The Laplace transform $g(s)$ satisfies

$$\left(\frac{e^{-sh} - e^{-s}}{s} + e^{-s}g(s) \right)^2 = -\frac{d}{ds}g(s). \qquad (4.5)$$

(ii) There is a constant $a(h) > 0$ such that

$$\lim_{x \to \infty} \frac{1}{x} \int_0^x e^{a(h)(u+1)} Pr(L(u) \geq h) du = 1. \qquad (4.6)$$

Proof. The Laplace transform of $xf(x+1)$ is $-\frac{d}{ds}g(s)$. By the convolution property of Laplace transform we have

$$\int_0^\infty e^{-sx} \left(\int_0^x f(x-y)f(y)dy \right) = \left(\int_0^\infty e^{-sx} f(x)dx \right)^2$$
$$= \left(\frac{e^{-sh} - e^{-s}}{s} + e^{-s}g(s) \right)^2.$$

Then Equation (4.5) follows from Equation (4.3) and the initial condition (4.4).

Put $\alpha(s) = g(-s)$. Then

$$\left(\frac{e^{sh} - e^s}{s} - e^s \alpha(s) \right)^2 = \frac{d}{ds}\alpha(s).$$

From Lemma 4.3, $\alpha(s)$ blows up at a certain point $a(h) > 0$. Hence

$$\frac{\alpha'(s)}{(\alpha(s))^2} \xrightarrow[s \to -0 + a(h)]{} e^{2a(h)}.$$

So, by integrating the equivalent we get:

$$\frac{-1}{\alpha(s)} \approx (s - a(h))e^{2a(h)} \quad \text{for} \quad s \to -0 + a(h).$$

Hence if $f_2(x) = e^{a(h)x} f(x+1)$ one has

$$L(f_2)(s) \approx \frac{e^{-2a(h)}}{s} \quad \text{for } s \ll 1.$$

So, by applying Theorem 3.2 we get the equivalent

$$\int_0^T e^{a(h)x} f(x+1)dx \simeq e^{-2a(h)} T \text{ for } T \to \infty.$$

After rewriting this implies the limit (4.6). □

We should point out that $a(h)$ is an exponential decay and that the function $f(x)$ itself has an oscillatory behavior. Equation (4.6) shows that the function $e^{-a(h)(x+1)}$ approximates well $f(x)$. Figure 4.1 shows more graphically how well $f(x)$ is approximated by this exponential decay.

Now consider

$$G(s) = \int_0^\infty e^{-sx} f(x)dx. \qquad (4.7)$$

It is easy to see that $G(s) = \frac{e^{-sh} - e^{-s}}{s} + e^{-s}g(s)$ and one has the following result:

Theorem 4.2. *There exists a holomorphic function H defined on $\mathbb{C}$ such that*

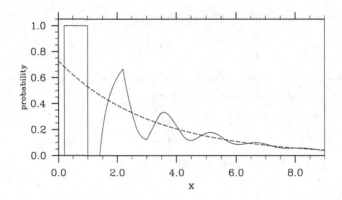

Fig. 4.1 The function $f(x)$ and its exponential approximation $\exp(-a(h)(x+1))$ for the case $h = 0.2$

(i) $G(s)$ converges for $\operatorname{Re} s > -a(h)$ and G has a pole at $-a(h)$.

(ii) the following relation holds

$$G(s) = \frac{H'(s)}{H(s)}. \tag{4.8}$$

(iii) H satisfies

$$H'' + 2H' = \left\{ \frac{(1-h)}{s} e^{-(1+h)s} - \frac{1}{s^2}[e^{-(1+h)s} - e^{-2s}] \right\} H. \tag{4.9}$$

(iv) H has only simple zeros.

(v) H has no zero in the half plane $\operatorname{Re} z > -a(h)$.

(vi) $-a(h)$ is the only real zero of H.

Proof.   Theorem 4.1 implies that the integral defining G converges if $\operatorname{Re} s > -a(h)$ and the singularity at $-a(h)$ is obvious from Equation (4.6). Equation (4.8) is rewritten as

$$G(s) = (\log H(s))'.$$

Thus by integrating G and taking the exponential we get the existence of H satisfying (4.8) on the half plane $\operatorname{Re} s > -a(h)$. The differential equation (4.9) follows from (4.5) by simple rewriting. This equation is linear and the coefficients are holomorphic on $\mathbb{C}$, i.e. the apparent singularity at 0 in the coefficient of Equation (4.9) is actually removable as can be proved by a Taylor expansion. As a consequence, the solutions of the equation are naturally defined over $\mathbb{C}$ and thus H also. H has only simple zeros because if z is a multiple zero of H then $H(z) = H'(z) = 0$, which implies $H = 0$ by the differential equation.

We have $|G(s)| \leq G(Re\, s)$ since f is positive. Thus G is finite if $Re\, s >$ $-a(h)$ and this implies that H is nonzero for $Re\, s > -a(h)$ that is (v) holds.

If one writes $H(s) = e^{-s}H_2(s)$ then Equation (4.9) is rewritten as

$$H_2''(s) = W_2(s)H_2(s)$$

with

$$W_2(s) = \frac{(1-h)}{s}e^{-(1+h)s} - \frac{1}{s^2}\left[e^{-(1+h)s} - e^{-2s}\right] + 1.$$

We can write

$$W_2(s) = 1 + \frac{e^{-2s}}{s^2}\left\{s(1-h)e^{s(1-h)} - e^{s(1-h)} + 1\right\}.$$

The function $ve^v - e^v + 1$ is positive on $\mathbb{R}$ and thus W_2 as well. Suppose now that H has two distinct real zeros says x_1 and x_2. Then the function H admits a local maximum in $x_0 \in (x_1, x_2)$, which implies that $H'(x_0) = 0$. But one has $H''(x_0) = W_2(x_0)H(x_0) > 0$, which implies that H has a local minimum at x_0, which is a contradiction and (vi) holds. $\qquad\square$

Note that the substitution $G = \frac{H'}{H}$ is the standard method for reducing *Ricatti's equations* that is equations of the form $y' = y^2 + Ry + Q$ to a second order linear differential equation. In our case this means that if H_1 and H_2 are two solutions which form a basis of the space of solutions, then the function G is expressed as

$$G = \frac{\alpha_1 H_1' + \alpha_2 H_2'}{\alpha_1 H_1 + \alpha_2 H_2}.$$

See among many possible references, [Ince (1944)].

4.3 Numerical calculations for $a(h)$

We give an algorithm to get a lower bound and an upper bound of $a(h)$. Since $0 \leq g(s) \leq \int_0^\infty e^{-sx}dx = 1/s$, we have

$$\lim_{s\to\infty} g(s) = 0.$$

Putting $t = e^{-2s}$, and defining $\phi(t)$ by $\phi(t) = g(s)$, we have

$$\begin{cases} 2\phi'(t) = [\phi(t) + C(t)]^2 \text{ with } \phi(0) = 0 \\ \text{and } C(t) = \dfrac{2(1 - t^{\frac{1}{2}(h-1)})}{\log t}. \end{cases} \qquad (4.10)$$

Since the function $G(s)$ is defined on $(-a(h), \infty)$, the function ϕ is defined on $[0, e^{2a(h)})$.

The difficulty of this equation is that while $C^2(t)$ is integrable at $t = 0$ for $h > 0$, $C(t)$ is not even continuous at $t = 0$. As a consequence, we cannot apply the classical Cauchy Lipschitz theorem as in [Cesari (1963)], p. 4. Thus, in order to prove existence and uniqueness of solution we have to adapt the proof given there to our case. In addition the proof will give us an explicit algorithm for getting upper bounds on ϕ.

Definition 4.1. If V is a vector space then a norm N on V is a function $N : V \to \mathbb{R}$ such that

(1) $N(x + y) \leq N(x) + N(y)$ for all $x, y \in V$.
(2) $N(\lambda x) = |\lambda| N(x)$ for $\lambda \in \mathbb{R}$ and $x \in V$.
(3) $N(x) = 0$ is equivalent to $x = 0$.

A sequence $(x_n)_{n \in \mathbb{N}}$ with $x_n \in V$ is called a *Cauchy sequence* if for all $\epsilon > 0$ there exists $n_0 \in \mathbb{N}$ such that

$$N(x_{n+m} - x_n) \leq \epsilon \text{ for every } n \geq n_0 \text{ and } m \in \mathbb{N}.$$

A sequence $(x_n)_{n \in \mathbb{N}}$ converges if there exists $x \in V$ (the limit of x_n) such that for all $\epsilon > 0$ there exists $n_0 \in \mathbb{N}$ such that

$$N(x_n - x) \leq \epsilon \text{ for every } n \geq n_0.$$

A normed space V is called a *Banach space* if all Cauchy sequences converge.

If we take the real numbers and put the absolute value as norm then it is a Banach space. The notion of Cauchy sequence is key to the theoretical definition of integrals, absolute convergence, etc. since it allows to prove the existence of some objects without actually computing them. The following fixed point theorem is a typical example:

Theorem 4.3. *If V is a Banach space, A is a subset of V and $K : A \to A$ is a function such that there exists $k < 1$ with*

$$N(K(x) - K(y)) \leq kN(x - y) \text{ for } x, y \in A$$

then there exist a unique $a \in A$ such that $K(a) = a$ and for all $x \in A$ the sequence $(x, K(x), K(K(x)), \dots)$ converges towards a.

Theorem 4.4. *Fix a $h > 0$, then there is a $\tau > 0$ such that Equation (4.10) admits a unique solution on $[0, \tau]$.*

Proof. Let us take $R > 0$, we want to prove that there is a $\tau > 0$ such that there is a solution to Equation (4.10) with $|\phi| \leq R$ on $[0, \tau]$. Define $\mathcal{F}_\tau$ to be the space of continuous functions on $[0, \tau]$. For the norm

$\|f\|_\infty = \sup_{x \in [0,\tau]} |f(x)|$ this is a Banach space. The differential Equation (4.10) is equivalent to the integral equation

$$\phi = K(\phi) \text{ with } K(f)(x) = \frac{1}{2} \int_0^x [f(t) + C(t)]^2 dt \text{ for } 0 \le x \le \tau.$$

We want to apply the fixed point theorem to K for some $\tau > 0$ to be defined later, which would prove the existence and uniqueness of a solution. We define $B(0, R)$ to be the set of functions $f \in \mathcal{F}_\tau$ bounded by R:

$$B(0, R) = \{f \in \mathcal{F}_\tau \text{ s.t. } \|f\|_\infty \le R\}.$$

We have if $f \in B(0, R)$ the bound

$$
\begin{aligned}
|K(f)(x)| &\le \tfrac{1}{2} \int_0^x f(t)^2 + 2|f(t)|C(t) + C(t)^2 dt \\
&\le \tfrac{1}{2} \int_0^\tau f(t)^2 + 2|f(t)|C(t) + C(t)^2 dt \\
&\le \tfrac{1}{2} \int_0^\tau R^2 + 2RC(t) + C(t)^2 dt \\
&\le \tfrac{\tau}{2} R^2 + R \int_0^\tau C(t) dt + \tfrac{1}{2} \int_0^\tau C(t)^2 dt.
\end{aligned}
$$

This relation is now expressed as

$$\|K(f)\|_\infty \le \frac{\tau}{2} R^2 + R \int_0^\tau C(t) dt + \frac{1}{2} \int_0^\tau C(t)^2 dt.$$

Since both integrals $\int_0^\tau C(t) dt$ and $\int_0^\tau C(t)^2 dt$ converge we can find a $\tau > 0$ such that $\|K(f)\|_\infty \le R$.

On the other hand, we have if $f_1, f_2 \in B(0, R)$ the inequalities

$$
\begin{aligned}
|K(f_1)(x) - K(f_2)(x)| &\le \tfrac{1}{2} \int_0^x |f_1(t)^2 - f_2(t)^2| + 2|f_1(t) - f_2(t)|C(t) dt \\
&\le \tfrac{1}{2} \|f_1 - f_2\|_\infty \int_0^x 2R + 2C(t) dt \\
&\le \|f_1 - f_2\|_\infty \left\{ R\tau + \int_0^\tau C(t) dt \right\}
\end{aligned}
$$

which imply

$$\|K(f_1) - K(f_2)\|_\infty \le \|f_1 - f_2\|_\infty \left\{ R\tau + \int_0^\tau C(t) dt \right\}.$$

So, we can ask that in addition $c = R\tau + \int_0^\tau C(t) dt < 1$, which implies $\|K(f_1) - K(f_2)\|_\infty \le c\|f_1 - f_2\|_\infty$. We can then apply the fixed point theorem, which gives existence and uniqueness of the solution. $\square$

After one has the existence on the interval $[0, \tau]$, we can apply the standard Cauchy Lipschitz theorem and get the existence of a maximal interval $[0, a(h)[$ on which ϕ exists. One interest of the analysis of Theorem 4.4 is that it gives explicitly computable bounds. If for some $h > 0$ there exists R, τ satisfying to

$$
\begin{cases}
\frac{\tau}{2} R^2 + R \int_0^\tau C(t) dt + \frac{1}{2} \int_0^\tau C(t)^2 dt \le R, \\
R\tau + \int_0^\tau C(t) dt < 1,
\end{cases}
\tag{4.11}
$$

then there exists a solution ϕ on $[0, \tau]$ with $|\phi(t)| \leq R$ hence $a(h) > \tau$. In order to estimate the integrals we use the bound

$$0 < \int_0^\tau C(t)^2 dt \leq \int_0^\tau \frac{t^{h-1}}{(\log t)^2} dt \leq \frac{1}{(\log \tau)^2} \int_0^\tau t^{h-1} dt \leq \frac{1}{(\log \tau)^2} \frac{\tau^h}{h}$$

and similarly for $\int_0^\tau C(t) dt$.

Due to the singularity at 0, we cannot apply standard methods like Runge Kutta to Equation (4.10) and we have to use some specific methods. The method, we used for estimating $a(h)$ uses monotonicity results for the coefficients of the equation. Assuming C is a constant, the differential equation is written as

$$\frac{d}{dt}\left(-\frac{1}{\phi(t) + C}\right) = \frac{1}{2}.$$

Suppose that $\underline{\phi}_1$ is a lower bound of ϕ at τ_1. Then $\underline{\phi}_2$ defined by

$$\frac{1}{\underline{\phi}_1 + C(\tau_2)} - \frac{1}{\underline{\phi}_2 + C(\tau_2)} = \frac{1}{2}(\tau_2 - \tau_1)$$

is a lower bound of ϕ at τ_2. Similarly, if $\overline{\phi}_1$ is an upper bound of ϕ on $[0, \tau_1]$ then $\overline{\phi}_2$ defined by

$$\frac{1}{\overline{\phi}_1 + C(\tau_1)} - \frac{1}{\overline{\phi}_2 + C(\tau_1)} = \frac{1}{2}(\tau_2 - \tau_1)$$

is an upper bound of ϕ on $[0, \tau_2]$. This is because on the interval $[\tau_1, \tau_2]$ we have the differential inequality

$$[\phi(t) + C(\tau_1)]^2 \geq 2\phi'(t) \geq [\phi(t) + C(\tau_2)]^2.$$

The method is then the following: we choose a τ_0, we derive an upper bound of ϕ on $[0, \tau_0]$ using (4.11) and we incrementally find upper bound of ϕ on $[0, \tau_k]$ with $\tau_k = \tau_0 + k\delta$ until we cannot do so and thus a lower bound T_{low} on $e^{2a(h)}$ is τ_{k-1}.

For the upper bound on $a(h)$, we fix the lower bound of 0 for ϕ at τ_0 and incrementally find better lower bound at τ_k until we find a negative or an infinite value, which means that τ_k is an upper bound T_{upp} of $e^{a(h)}$.

Using the relation $t = e^{-2s}$, we can obtain the bounds for $a(h)$, that is to say, $\frac{1}{2} \log T_{low}$ is a lower bound and $\frac{1}{2} \log T_{upp}$ is an upper bound for $a(h)$. The lower and upper bounds are shown in Figure 4.2 and the numerical values are given in Table 4.1. Note that if one is content with some approximate computation with no error analysis then a simpler computation method is to approximate $f(x) = Pr(L(x) \geq h)$ by directly integrating Equation (4.3) and then use a regression analysis for computing the decay coefficient. Figure 4.1 shows that the probability converge $f(x)$ fits very well its asymptotic limit.

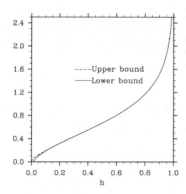

Fig. 4.2   Lower and Upper bounds for $a(h)$

4.4   Asymptotic analysis for $a(h)$

In this section we use singular perturbation techniques to investigate the values of $a(h)$ when h is small, and when h is close to 1. We use the Laplace transform and as before we write

$$G(s) = \frac{H'(s)}{H(s)}.$$

Then by Theorem 4.2 the function H satisfies

$$
\begin{aligned}
H'' + 2H' &= \left\{ \frac{(1-h)}{s} e^{-(1+h)s} - \frac{1}{s^2} [e^{-(1+h)s} - e^{-2s}] \right\} H \\
&= \frac{e^{-2s}}{s^2} \left\{ s(1-h)e^{s(1-h)} - e^{s(1-h)} + 1 \right\} H \\
&= (1-h)^2 e^{-2s} \psi(s(1-h)) H
\end{aligned}
\tag{4.12}
$$

with $\psi(u) = \frac{ue^u - e^u + 1}{u^2}$ a function holomorphic over $\mathbb{C}$.

In Subsection 4.4.1 we consider the Renewal Theory method that allows to find the first order of the expansion of $a(h)$ when $h \to 1$. In Subsection 4.4.2, following [Morrison (1987)], we determine the full asymptotic expansion of $a(h)$ in the case $h \to 1$, by using the above differential equation. In Subsection 4.4.3 we consider the case $0 < h \ll 1$ again by following [Morrison (1987)]. This section is adapted from [Morrison (1987)], we tried to complement the argument exposed there by giving some additional details that makes the proof hopefully clearer. For that we have to introduce some relatively advanced tools of asymptotic analysis. Furthermore, in order to simplify the exposition, we limit ourselves to the first terms of the perturbation expansion instead of [Morrison (1987)], which gives the second order terms. Also [Morrison (1987)] showed in pictures that his expansion fits the obtained values of $a(h)$ quite well for $0 < h \ll 1$.

Table 4.1 Lower and upper bounds for $a(h)$ as a function of h

h	lower	upper	h	lower	upper	h	lower	upper
0.01	-0.03742	0.05659	0.34	0.47976	0.48004	0.67	0.90843	0.90843
0.02	0.02331	0.07288	0.35	0.49123	0.49147	0.68	0.92510	0.92511
0.03	0.05576	0.08876	0.36	0.50270	0.50291	0.69	0.94220	0.94220
0.04	0.08018	0.10427	0.37	0.51420	0.51438	0.70	0.95976	0.95976
0.05	0.10098	0.11942	0.38	0.52572	0.52587	0.71	0.97780	0.97780
0.06	0.11972	0.13424	0.39	0.53727	0.53741	0.72	0.99637	0.99637
0.07	0.13712	0.14876	0.40	0.54887	0.54898	0.73	1.01551	1.01551
0.08	0.15354	0.16299	0.41	0.56051	0.56061	0.74	1.03527	1.03527
0.09	0.16922	0.17695	0.42	0.57220	0.57229	0.75	1.05570	1.05570
0.10	0.18496	0.19005	0.43	0.58396	0.58404	0.76	1.07686	1.07686
0.11	0.19945	0.20362	0.44	0.59579	0.59586	0.77	1.09882	1.09882
0.12	0.21354	0.21696	0.45	0.60770	0.60775	0.78	1.12165	1.12165
0.13	0.22728	0.23010	0.46	0.61968	0.61973	0.79	1.14544	1.14544
0.14	0.24072	0.24305	0.47	0.63177	0.63181	0.80	1.17028	1.17028
0.15	0.25389	0.25582	0.48	0.64395	0.64399	0.81	1.19630	1.19630
0.16	0.26683	0.26843	0.49	0.65624	0.65627	0.82	1.22363	1.22363
0.17	0.27956	0.28089	0.50	0.66865	0.66867	0.83	1.25242	1.25242
0.18	0.29211	0.29321	0.51	0.68118	0.68120	0.84	1.28285	1.28285
0.19	0.30449	0.30541	0.52	0.69385	0.69387	0.85	1.31514	1.31514
0.20	0.31596	0.31816	0.53	0.70667	0.70668	0.86	1.34957	1.34957
0.21	0.32816	0.33006	0.54	0.71964	0.71965	0.87	1.38646	1.38646
0.22	0.34024	0.34187	0.55	0.73278	0.73279	0.88	1.42620	1.42620
0.23	0.35221	0.35361	0.56	0.74609	0.74610	0.89	1.46931	1.46932
0.24	0.36408	0.36529	0.57	0.75960	0.75961	0.90	1.51645	1.51645
0.25	0.37586	0.37691	0.58	0.77331	0.77332	0.91	1.56849	1.56849
0.26	0.38757	0.38848	0.59	0.78724	0.78724	0.92	1.62658	1.62658
0.27	0.39922	0.40000	0.60	0.80139	0.80140	0.93	1.69239	1.69239
0.28	0.41081	0.41149	0.61	0.81580	0.81580	0.94	1.76832	1.76832
0.29	0.42237	0.42295	0.62	0.83047	0.83047	0.95	1.85812	1.85812
0.30	0.43388	0.43439	0.63	0.84542	0.84542	0.96	1.96807	1.96807
0.31	0.44537	0.44581	0.64	0.86066	0.86067	0.97	2.10998	2.10998
0.32	0.45684	0.45722	0.65	0.87623	0.87624	0.98	2.31035	2.31035
0.33	0.46830	0.46863	0.66	0.89215	0.89215	0.99	2.65391	2.65391

4.4.1 Renewal equation technique

We will derive later using analytical tools some quite precise estimation of $a(h)$ both for small h and h near 1 but before that we indicate some estimation of $a(h)$ when h is around 1.

Lemma 4.4. *We have the following inequalities:*

$$\frac{(1-h)^k}{2^{k-1}}Q_h^{k\star}(x+1) \le f_k(x) \le \frac{(1-h)^k}{(2h)^{k-1}}Q_h^{k\star}(x+1)$$

where, $Q_h(x) = P_h(x-1)$. *Namely,*

$$Q_h(x) = \begin{cases} \dfrac{1}{1-h} & \text{for } h+1 \le x < 2, \\ 0 & \text{for } x < h+1, \\ 0 & \text{for } 2 \le x. \end{cases}$$

Proof. The proof is similar to the one of Lemma 4.2, the right-hand side being a direct translation of it. □

For $0 < h < 1$, let us define

$$f_{l,h}(x) = \sum_{k=1}^{\infty} \frac{(1-h)^k}{2^k} Q_h^{k\star}(x)$$

and

$$f_{u,h}(x) = \sum_{k=1}^{\infty} \frac{(1-h)^k}{(2h)^k} Q_h^{k\star}(x).$$

Lemma 4.5. *(i) The functions $f_{l,h}$ and $f_{u,h}$ satisfy*

$$2f_{l,h}(x) \le f(x-1) \le 2hf_{u,h}(x) \quad \text{for } 0 \le h \le 1.$$

(ii) We have the inequality

$$a(h,u) \le a(h) \le a(h,l)$$

with $a(h,l)$ and $a(h,u)$ the unique strictly positive solutions of

$$\begin{cases} 2a(h,l) = e^{2a(h,l)} - e^{(h+1)a(h,l)} \\ 2ha(h,u) = e^{2a(h,u)} - e^{(h+1)a(h,u)}. \end{cases}$$

Proof. Inequality (i) follows from Lemma 4.4 by summation. Henceforth, we apply Theorem 4.10 from the theory of renewal processes to the above $f_{l,h}(x)$ and $f_{u,h}(x)$. First, we consider the function $f_{l,h}(x)$. We have the following integral

$$\int_0^{\infty} e^{ky} \frac{1-h}{2} Q_h(y) dy = \frac{1-h}{2} \int_{h+1}^2 \frac{1}{1-h} e^{ky} dy$$
$$= \frac{1}{2}\frac{1}{k} \left[e^{2k} - e^{(1+h)k} \right].$$

Thus we obtain that $a(h,l)$ is the only positive real satisfying

$$\int_0^{\infty} e^{a(h,l)y} \frac{1-h}{2} Q_h(y) dy = 1.$$

We are able to define a proper density $L_{l,h}^{\sharp}(x)$ by

$$L_{l,h}^{\sharp}(x) = e^{a(h,l)x} \frac{1-h}{2} Q_h(x).$$

We define the function $f_{l,h}^{\sharp}(x)$ for the above constant $a(h, l)$ as follows:

$$f_{l,h}^{\sharp}(x) = e^{a(h,l)x} f_{l,h}(x).$$

Similarly $a(h, u)$ is the unique positive real such that

$$\int_0^\infty e^{a(h,u)y} \frac{1-h}{2h} Q_h(y) dy = 1.$$

We are also able to define proper densities $L_{u,h}^{\sharp}(x)$ by

$$L_{u,h}^{\sharp}(x) = e^{a(h,l)x} \frac{1-h}{2h} Q_h(x).$$

Furthermore, by the similar definition as above $f_{l,h}^{\sharp}(x)$, we also define the functions $f_{u,h}^{\sharp}(x)$ as follows:

$$f_{u,h}^{\sharp}(x) = e^{a(h,u)x} f_{u,h}(x)$$

which is as we need.

Now the application of the renewal Theorem 4.10 gives the convergence

$$f_{l,h}^{\sharp}(x) \to \frac{1}{\mu_{l,h}^{\sharp}} \quad \text{and} \quad f_{u,h}^{\sharp}(x) \to \frac{1}{\mu_{u,h}^{\sharp}}$$

as x tends to infinity, where $\mu_{l,h}^{\sharp}$, and $\mu_{u,h}^{\sharp}$ are expectations of $L_{l,h}^{\sharp}(x)$ and $L_{u,h}^{\sharp}(x)$ respectively as

$$\mu_{l,h}^{\sharp} = \frac{e^{2a(h,l)}}{2(a(h,l))^2} \left\{ 2\,a(h,l) - 1 - ((h+1)a(h,l) - 1)\,e^{-a(h,l)(1-h)} \right\}$$

and

$$\mu_{u,h}^{\sharp} = \frac{e^{2a(h,u)}}{2h(a(h,u))^2} \left\{ 2a(h,u) - 1 - ((h+1)a(h,u) - 1)\,e^{-a(h,u)(1-h)} \right\}.$$

Thus $f_{l,h}$ and $f_{u,h}$ have exponential decrease $a(h,l)$ and $a(h,u)$ and so this obviously implies (ii). ☐

Theorem 4.5. *We have*

$$a(h) \simeq \frac{1}{2} \log \left(\frac{2}{1-h} \right)$$

as h goes to 1.

Proof. We are considering first the asymptotic behavior of $a(h, u)$ and $a(h, l)$ as h goes to 1. One gets

$$\begin{cases} a(h,l) \cong \frac{1}{2} \log \frac{2}{1-h}, \\ a(h,u) \cong \frac{1}{2} \log \frac{2h}{1-h} \end{cases}$$

and thus the required asymptotic behavior follows. ☐

4.4.2   *Approximation for small $1 - h$*

In this section we will investigate the solution of (4.9), subject to the asymptotic behavior of (4.13), when $h = 1 - \epsilon$ and $0 < \epsilon \ll 1$.

Lemma 4.6. *(i) We have $f(x) = 0$ if $x \notin J_\epsilon$ with*

$$J_\epsilon = \bigcup_{k \geq 0} [2k + 1 - (k + 1)\epsilon, 2k + 1).$$

(ii) On $[2k + 1 - (k + 1)\epsilon, 2k + 1)$ we have $f(x) = O(\epsilon^k)$.

Proof.   (i) is clear for $k = 0$ and for $k = 1$ it follows from Equation (4.3). The general case is by induction. (ii) is proved similarly by remarking that the length of the interval decreases with ϵ.   □

We should remark that for any $\epsilon > 0$ the set J_ϵ is the union of a finite number of intervals of the form $[2k + 1 - (k + 1)\epsilon, 2k + 1)$ and an infinite interval $[a, \infty[$.

We will now use those estimates for the functions $G(s)$ and $H(s)$. We expect $H(s)$ to have a zero around $-a_0$ where

$$a_0 = \frac{1}{2} \log(2/\epsilon).$$

Thus we set

$$s = \xi - a_0 \text{ and } H(s) = K(\xi).$$

Lemma 4.7. *(i) For $s \gg 1$ we have*

$$\frac{H'(s)}{H(s)} = G(s) = \frac{e^{-s}}{s}\left[(e^{-hs} - e^{-s}) + O\left(e^{-(1+2h)s}\right)\right]. \tag{4.13}$$

(ii) For $\xi \gg 1$ and $\epsilon\xi \ll 1$ we have

$$\begin{aligned}
\frac{K'(\xi)}{K(\xi)} &= 2e^{-2\xi}\left[1 + \frac{\epsilon(\xi - a_0)}{2} + \frac{\epsilon^2(\xi - a_0)^2}{6} + O\left(\epsilon^3(\xi - a_0)^3\right)\right] \\
&\quad + O(e^{-4\xi}).
\end{aligned} \tag{4.14}$$

Proof.   From Equations (4.4) and (4.7) we get

$$G(s) = \int_h^1 e^{-s(x+1)}dx + \int_{1+2h}^\infty e^{-s(x+1)}f(x)dx.$$

But, from (4.2), it follows that $0 \leq f(x) \leq 1$ for $x \geq 0$. Hence

$$0 \leq G(s) - \frac{1}{s}e^{-s}(e^{-hs} - e^{-s}) \leq \frac{1}{s}e^{-2(1+h)s}.$$

Consequently (i) follows. By using the expression of $K(\xi)$ and substituting we get

$$\frac{K'(\xi)}{K(\xi)} = e^{-2s}\frac{e^{\epsilon s}-1}{s} + O\left(e^{-(4-2\epsilon)s}\right)$$
$$= 2e^{-2\xi}\left[1 + \frac{\epsilon(\xi-a_0)}{2} + \frac{\epsilon^2(\xi-a_0)^2}{6} + O\left(\epsilon^3(\xi-a_0)^3\right)\right] + O(e^{-4\xi}).$$

We used Lemma 4.6 to remove the ϵ term from the equation. $\qquad\square$

Now we have one variable ϵ but we have to deal with a logarithmic singularity. Hence we write $\delta = \epsilon a_0$ and we will use a two-variable implicit theorem for the computation [Avez (1986)].

For $\epsilon|s| \ll 1$ we obtain from (4.12)

$$H'' + 2H' = \epsilon^2 e^{-2s}\left[\frac{1}{2} + \frac{\epsilon s}{3} + O(\epsilon^2 s^2)\right]H. \qquad (4.15)$$

Theorem 4.6. *[Morrison (1987)] (i) For any k we have an expansion*

$$a(h) = a_0 + \sum_{l=1}^{k} \epsilon^l P_l(a_0) + O\left(\epsilon^{k+1}\log(\epsilon)^{k+1}\right)$$

with P_l a polynomial of degree l.
(ii) We have

$$a(h) = a_0 + \frac{\epsilon}{4}\left(a_0 - \frac{3}{4}\right) + O\left(\epsilon^2\log(\epsilon)^2\right).$$

Proof. From (4.15), we obtain

$$K'' + 2K' = 2\epsilon e^{-2\xi}\psi(\epsilon(\xi - a_0))K$$
$$= 2\epsilon e^{-2\xi}\psi(\epsilon\xi - \delta)K \qquad (4.16)$$
$$= 2\epsilon e^{-2\xi}\left[1 + \tfrac{2}{3}(\epsilon\xi - \delta) + O\left((\epsilon\xi - \delta)^2\right)\right]K.$$

Equation (4.16) has an analytic dependency on ϵ and δ. Thus the solution H can be expanded as

$$K(\xi) = \sum_{m=0}^{\infty}\left(\sum_{i+j=m} \epsilon^i \delta^j K_{i,j}(\xi)\right). \qquad (4.17)$$

We also write

$$K_m(\xi) = \sum_{i+j=m} \epsilon^i \delta^j K_{i,j}(\xi).$$

By substitution into (4.16) we find that

$$K_0'' + 2K_0' = 0 \text{ and } K_1'' + 2K_1' = \epsilon e^{-2\xi}K_0.$$

The general expression of K_0 is

$$K_0 = \alpha_0 + \beta_0 e^{-2\xi} = \alpha_0 + \frac{\epsilon}{2}\beta_0 e^{2s},$$

where α_0 and β_0 are constants. The asymptotic behavior for $s \gg 1$ and $\epsilon s \ll 1$, corresponding to (4.14) and (4.17), implies that $\beta_0 = -\alpha_0$, and we may take $\alpha_0 = 1$ without loss of generality, since K is undetermined to within a multiplicative constant. Hence,

$$K_0 = 1 - e^{-2\xi}$$

and we have

$$(e^{2\xi}K_1')' = \epsilon(1 - e^{-2\xi}),$$

which implies that

$$K_1 = \alpha_1 + \left(\beta_1 - \frac{\epsilon\xi}{2}\right)e^{-2\xi} - \frac{\epsilon}{8}e^{-4\xi}.$$

We may take $\alpha_1 = 0$ without loss of generality. Then, from the asymptotic behavior for $s \gg 1$ and $\epsilon s \ll 1$, we obtain $\beta_1 = \frac{1}{2}(\delta - \frac{\epsilon}{2})$, so that

$$K_1 = \frac{1}{2}\left(\delta - \frac{\epsilon}{2} - \epsilon\xi\right)e^{-2\xi} - \frac{\epsilon}{8}e^{-4\xi}.$$

This can obviously be extended to all degrees and thus we can get an expression for $K_m(\xi)$ in the form of

$$K_m(\xi) = \sum_{k=0}^{m}\left\{\sum_{l=0}^{m-k} c_{m,k,l}\epsilon^{m-l}(\epsilon\xi - \delta)^l\right\}e^{-2(k+1)\xi}.$$

The function K_0 has a simple zero at $\xi = 0$. Thus we can apply the implicit function theorem [Avez (1986)] to prove the existence of a zero $a(\epsilon, \delta)$ of $K(\xi)$ that depends on ϵ and δ.

This function $a(\epsilon, \delta)$ is analytic on ϵ and δ since K is itself analytic and thus we get that it admits an expansion

$$a(\epsilon, \delta) = \sum_{k=1}^{m}Q_k(\epsilon, \delta) + O\left(\epsilon^{m+1} + \delta^{m+1}\right),$$

with Q_k a homogeneous polynomial of ϵ and δ. So, we get (i) by the following substitution:

$$a(h) = a(1 - \epsilon) = a_0 + a(\epsilon, \delta) = a_0 + a(\epsilon, \epsilon a_0).$$

(ii) follows by substituting $a(\epsilon, \delta) = A\epsilon + B\delta + O\left(\epsilon^2 + \delta^2\right)$. $\square$

In principle it is thus possible to obtain the expansion of $a(h)$ for $1 - h$ small at any order by using a computer.

4.4.3   *Approximation for small h*

In this section we will investigate the solution of (4.12) when $0 < h \ll 1$.
A special function named exponential integral is used in the derivation, see
[Magnus, Oberhettinger and Soni (1966); Abramowitz and Stegun (1965)]
for details on this special function:

Definition 4.2. For any $x \in \mathbb{R}_+^*$ we define

$$E_1(x) = \int_x^\infty \frac{e^{-t}}{t} \, dt.$$

For the function E_1 it holds:

(i) We have the asymptotic expansions:

$$\begin{cases} E_1(x) = O\left(\frac{e^{-x}}{x}\right) & \text{for } x \to \infty, \\ E_1(x) = -\gamma - \ln(x) + x + O(x^2) & \text{for } x \to 0^+ \end{cases}$$

with γ the Euler constant.

(ii) It admits an analytic continuation to $\mathbb{C} - \mathbb{R}_-$.

(iii) The function $\exp(E_1(t))$ has an analytic continuation to $\mathbb{C} - \{0\}$.

We should note that the usual definition of E_1 is different with a minus
sign added to x, but we find it more convenient to work with this notation.
    Before giving the theorem, we need to expose a notion of singularity for
ordinary differential equations, that of *regular singularity*:

Definition 4.3. (i) If $a_1(z)$, $a_2(z)$, $\ldots$, $a_n(z)$ are meromorphic functions
on a domain $D \subset \mathbb{C}$ then we say that the ordinary differential

$$y^{(n)} + a_1(z)y^{(n-1)} + \cdots + a_n y = 0$$

has a *regular singularity* at $z_0 \in D$ if there exists a neighborhood $\Omega \subset D$
with $z_0 \in \Omega$ such that

$$a_i(z)(z - z_0)^i$$

are holomorphic on Ω for all $1 \le i \le n$.

    (ii) If $A(z)$ is a meromorphic function with matrix values on a domain
D of $\mathbb{C}$ then we say that the vector differential equation

$$v' = A(z)v$$

has a *regular singularity* at $z_0 \in D$ if there exist a neighborhood $\Omega \subset D$
with $z_0 \in D$ such that $(z - z_0)A(z)$ is holomorphic on Ω.

If y is a solution of an ordinary differential equation with a regular singularity at z_0 then we write

$$v(z) = (y(z), (z - z_0)y'(z), \ldots, (z - z_0)^{n-1}y^{(n-1)}(z))$$

and it is easy to see that the vector function $v(z)$ is a solution of $v' = A(z)v$ with A having a regular singularity at z_0. So, we can expose the theory in the context of regular singularity for the matrix case only. For further details we refer, for example, to [Ince (1944); Balser (1999); Deligne (1970); Coddington and Levinson (1955)]. The basic result of the theory is the following:

Theorem 4.7. *Let* $v' = A(z)v$ *be a vector differential equation of order* n *on* Ω *with a regular singularity only at* z_0. *Let us write*

$$A(z) = \frac{1}{z - z_0}A_{-1} + A_0 + \sum_{k=1}^{\infty}(z - z_0)^k A_k.$$

Then there exists $r > 0$ *such that on* $\Omega = D(z_0, r) - [z_0, x]$ *with* $|z_0 - x| = r$ *(that is a disk of radius* r *centered at* z_0 *minus an interval going from the boundary to* z_0*) a basis of solution is formed by*

$$Y(z) = B(z)(z - z_0)^{A_{-1}}$$

with B *an* $n \times n$ *matrix-valued function holomorphic on* $D(z_0, r)$ *and* $B(z_0) = I_n$.

One proof method is based on working out the expansion and proving that it converges. We cannot use the neighborhood $D(z_0, r) - \{z_0\}$ since it is not simply connected and thus *a priori* the term $(z - z_0)^{A_{-1}}$ could be multiply valued and thus not a function.

The matrix power $(z - z_0)^{A_{-1}}$ can be rewritten as

$$(z - z_0)^{A_{-1}} = \exp(A_{-1}\log(z - z_0))$$

with exp being the matrix exponential, which we explain below.

Definition 4.4. If A is an $n \times n$ matrix then we define the matrix exponential to be

$$\exp(A) = \sum_{k=0}^{\infty}\frac{1}{k!}A^k.$$

The matrix exponential satisfies the following properties:

(i) If A is a diagonal matrix

$$A = \begin{pmatrix} \lambda_1 & 0 & \dots & 0 \\ 0 & \lambda_2 & 0 & \vdots \\ \vdots & \ddots & \ddots & 0 \\ 0 & \dots & 0 & \lambda_n \end{pmatrix} \quad \text{then} \quad \exp(A) = \begin{pmatrix} e^{\lambda_1} & 0 & \dots & 0 \\ 0 & e^{\lambda_2} & 0 & \vdots \\ \vdots & \ddots & \ddots & 0 \\ 0 & \dots & 0 & e^{\lambda_n} \end{pmatrix}.$$

(ii) If P is an invertible $n \times n$ matrix then

$$\exp(P^{-1}AP) = P^{-1}\exp(A)P.$$

(iii) If A and B are two commuting $n \times n$ matrices then

$$\exp(A + B) = \exp(A)\exp(B).$$

(iv) If A is a nilpotent matrix, i.e. $A^{n_0} = 0$ for some integer n_0 then

$$\exp(A) = \sum_{k=0}^{n_0-1} \frac{1}{k!}A^k.$$

If a matrix A is diagonalizable then properties (i) and (ii) allow to compute the exponential efficiently: first diagonalize A by computing P, compute the exponential of the diagonal matrix and then the product.

If A is not diagonalizable then the computation is a little more complicated. One way is to use Dunford decomposition, i.e. decompose A as a sum of two matrices D and N with D diagonalizable, N nilpotent and $DN = ND$. Then use properties (iii) and (iv). This decomposition is also called Jordan decomposition and for 2×2 matrices it means that there exists an invertible P such that $P^{-1}AP$ is either

$$\begin{pmatrix} \lambda_1 & 0 \\ 0 & \lambda_2 \end{pmatrix} \quad \text{or} \quad \begin{pmatrix} \lambda & c \\ 0 & \lambda \end{pmatrix}$$

whose matrix exponentials are

$$\begin{pmatrix} e^{\lambda_1} & 0 \\ 0 & e^{\lambda_2} \end{pmatrix} \quad \text{or} \quad e^{\lambda}\begin{pmatrix} 1 & c \\ 0 & 1 \end{pmatrix}.$$

For our problem this means that if A_{-1} is diagonalized with eigenvalues $\alpha_1, \dots, \alpha_n$ then the solution is expressed in terms of $(z - z_0)^{\alpha_1}, \dots, (z - z_0)^{\alpha_n}$. If A_{-1} is not diagonalizable then the solution is obtained in terms of $(z - z_0)^{\alpha_i}(\log(z - z_0))^k$ with $0 \le k \le m_i - 1$ with m_i the multiplicity of α_i in the minimal polynomial of A_{-1}. Ordinary singular equations with regular singularities occur very frequently in mathematical physics and geometry; for example, the hypergeometric functions and Bessel functions are solutions of such equations. The interest of functions solution of equations with regular singularities is that they are not too singular, i.e. the term $(z - z_0)^{A-1}$ is the only difficulty in dealing with them.

Lemma 4.8. *(i) We have the values*

$$
Pr(L(x) \geq h) = \begin{cases} 0 & \text{if } x < h, \\ 1 & \text{if } h \leq x < 1, \\ 0 & \text{if } 1 \leq x \leq 1 + 2h, \\ 1 - \frac{2h}{x-1} & \text{if } 1 + 2h \leq x \leq 2 + h. \end{cases}
$$

(ii) If $s \gg 1$ then we have the following expansion of G:

$$
G(s) = \frac{e^{-s}}{s} \left\{ e^{-hs} + e^{-s} \left[e^{-2hs} - 1 + 2hs \, E_1(2hs) + O(e^{-(1+h)s}) \right] \right\}.
$$

(iii) If $s \gg 1$ and $hs \ll 1$ then we have

$$
G(s) = \frac{1}{s} e^{-(1+h)s} + e^{-2s} \Big\{ 2h \left[\gamma - 1 + \log(2hs) - hs + O(h^2 s^2) \right]
$$

$$
+ O\left(\frac{1}{s} e^{-(1+h)s} \right) \Big\}.
$$

Proof. Statement (i) follows by direct integration of Equation (4.3) and using initial condition (4.4). For example one has

$$
f(x+1) = \frac{1}{x} \int_h^{x-h} dy = 1 - \frac{2h}{x} \text{ for } 2h < x < 1 + h.
$$

Consequently, from (4.7), we have

$$
G(s) = \int_h^1 e^{-s(x+1)} dx + \int_{2h}^{1+h} e^{-s(x+2)} (1 - 2h/x) dx + \int_{2+h}^{\infty} e^{-s(x+1)} f(x) dx. \tag{4.18}
$$

But, in term of the exponential integral

$$
\int_{2h}^{1+h} e^{-sx} \frac{dx}{x} = E_1(2hs) - E_1((1+h)s). \tag{4.19}
$$

Since $0 \leq f(x) \leq 1$ for $x \geq 0$, it follows from (4.18), (4.19) that

$$
G(s) = \frac{e^{-s}}{s} \left\{ e^{-hs} + e^{-s} \left[e^{-2hs} - 1 + 2hs \, E_1(2hs) + O(e^{-(1+h)s}) \right] \right\},
$$

for $s \gg 1$. By using the expansion of E_1 we have for $s \gg 1$ and $hs \ll 1$ the estimation

$$
G(s) = \frac{1}{s} e^{-(1+h)s} + e^{-2s} \Big\{ 2h \left[\gamma - 1 + \log(2hs) - hs + O(h^2 s^2) \right]
$$

$$
+ O\left(\frac{1}{s} e^{-(1+h)s} \right) \Big\},
$$

which is the required result.   □

Theorem 4.8. *[Morrison (1987)] We have*

$$a(h) = -e^{-2\gamma}2h[\gamma - 1 + \log(2h) + 2I] + O\left(h^2(\log(h))^2\right) \ for \ h \to 0$$

with γ the Euler constant and $I = \int_0^\infty \frac{\log t}{t}e^{-t}\,e^{-2E_1(t)}\,dt$.

Proof. If we write

$$G = \frac{1}{s}e^{-(1+h)s} + \frac{U'}{U},$$

then Equation (4.12) is rewritten as

$$U'' + 2\left[1 + \frac{1}{s}e^{-(1+h)s}\right]U' = \frac{e^{-2s}}{s^2}(1 - e^{-2hs})U \qquad (4.20)$$

and Lemma 4.8 (iii) implies for $s \gg 1$ and $hs \ll 1$ the estimation

$$\frac{U'(s)}{U(s)} = e^{-2s}\left\{2h\left[\gamma - 1 + \log(2hs) - hs + O(h^2s^2)\right]\right.$$
$$\left. + O\left(\frac{1}{s}\,e^{-(1+h)s}\right)\right\}. \qquad (4.21)$$

Equation (4.20) is a differential equation with a regular singularity at 0. Put $v_1 = U$ and $v_2 = sU'$. Equation (4.20) is written as

$$\left(\frac{v_2}{s}\right)' + 2\left[1 + \frac{1}{s}e^{-(1+h)s}\right]\frac{v_2}{s} = \frac{e^{-2s}}{s^2}(1 - e^{-2hs})v_1.$$

Hence we have

$$\begin{cases} v_1' = U' = \frac{1}{s}v_2, \\ v_2' = \frac{1}{s}v_2 + \frac{e^{-2s}}{s}(1 - e^{-2hs})v_1 - 2\left[1 + \frac{1}{s}e^{-(1+h)s}\right]v_2. \end{cases}$$

By expanding the coefficients we get

$$\begin{pmatrix} v_1' \\ v_2' \end{pmatrix} = A(s)\begin{pmatrix} v_1 \\ v_2 \end{pmatrix} \text{ with } A(s) = \frac{1}{s}\begin{pmatrix} 0 & 1 \\ 0 & -1 \end{pmatrix} + O(1).$$

Thus we get $A_{-1} = \begin{pmatrix} 0 & 1 \\ 0 & -1 \end{pmatrix}$, which is a diagonalizable matrix of eigenvalues 0 and -1. So, by Theorem 4.7 the solution is expressed in terms of a function behaving like $O(s^{-1})$ around 0 and a solution behaving like $O(1)$ around 0. No logarithm occur because the eigenvalues are simple. In other words, we know that we can find two basic solutions for U with the following expansion at 0:

$$\begin{cases} U_1(s,h) = \frac{1}{s} + O(s), \\ U_2(s,h) = 1 + O(s). \end{cases} \qquad (4.22)$$

One more direct method is to substitute $U = \frac{1}{s}V$ in (4.20) and get the following equation:

$$V'' + \left[2 + \frac{2}{s}e^{-(1+h)s} - \frac{2}{s}\right]V'$$

$$+ \left[\frac{2}{s^2} - \frac{2}{s} - \frac{2}{s^2}e^{-(1+h)s} - \frac{e^{-2s}}{s^2}(1 - e^{-2hs})\right]V = 0.$$

By doing expansions of the coefficients, one sees that they are actually analytic around 0. Thus we can apply standard tools and get a basis of solutions for this equation and then a basis of solution for Equation (4.20). One delicate point is that there is a dependency in h but this is not a problem since the dependency is holomorphic in h and we know that the proof of existence depends on power series arguments. So, there exist expansions

$$U_i(s,h) = U_{i,0}(s) + hU_{i,1}(s) + \sum_{k=2}^{\infty} h^k U_{i,k}(s) \text{ for } i = 1, 2 \qquad (4.23)$$

with $U_{i,k}(s)$ being functions holomorphic at 0 and thus on $\mathbb{C}$ since 0 is the only singularity of Equation (4.20). We are looking for the first two terms $U_{i,0}(s)$ and $U_{i,1}(s)$.

The functions $U_{i,0}(s)$ are solutions of the differential equation

$$y'' + 2\left[1 + \frac{1}{s}e^{-s}\right]y' = 0. \qquad (4.24)$$

We define

$$\Omega(s) = 2\left(s - \int_s^{\infty} e^{-t}\frac{dt}{t}\right) = 2[s - E_1(s)].$$

Then (4.24) may be written in the form

$$\left(e^{\Omega}y'\right)' = 0.$$

The solution is

$$y(s) = B_0 + A_0 \int_s^{\infty} e^{-\Omega(t)}dt,$$

where A_0 and B_0 are constants. We have the following equivalence at 0:

$$\Omega(t) = 2\gamma + 2\log(t) + O(t^2) \quad \text{and} \quad e^{-\Omega(t)} = \frac{e^{-2\gamma}}{t^2} + O(1).$$

From (4.22) and (4.23) we then have

$$U_{1,0}(s) = e^{2\gamma}\int_s^{\infty} e^{-\Omega(t)}dt \quad \text{and} \quad U_{2,0}(s) = 1.$$

Now expanding (4.20) in h and replacing we get

$$U_{i,1}'' + 2\left[1 + \frac{1}{s}e^{-s}\right]U_{i,1}' = \frac{2}{s}e^{-2s}U_{i,0} + 2e^{-s}U_{i,0}' \text{ for } i = 1, 2.$$

The expression of $U_{i,0}$ imply

$$\begin{cases} (e^{\Omega}U_{1,1}')' = e^{2\gamma}\frac{2}{s}e^{-2E_1(s)}\int_s^\infty e^{-\Omega(t)}dt - 2e^{2\gamma - s}, \\ (e^{\Omega}U_{2,1}')' = \frac{2}{s}e^{-2E_1(s)}. \end{cases}$$

Their solution is of the form

$$\begin{cases} U_{1,1}(s) = A_1 - B_1\int_s^\infty e^{-\Omega(t)}dt \\ \qquad + e^{2\gamma}\int_0^s e^{-\Omega(t)}\left\{\int_0^t \frac{2}{u}e^{-2E_1(u)}\int_u^\infty e^{-\Omega(v)}dv - 2e^{-u}du\right\}dt, \\ U_{2,1}(s) = A_2 - B_2\int_s^\infty e^{-\Omega(t)}dt + \int_0^s e^{-\Omega(t)}\left\{\int_0^t \frac{2}{u}e^{-2E_1(u)}du\right\}dt. \end{cases}$$

The terms not involving A_i, B_i are $O(s)$ thus in order to satisfy (4.23) and (4.22), we should have $A_i = B_i = 0$ and we get

$$\begin{cases} U_{1,1}(s) = e^{2\gamma}\int_0^s e^{-\Omega(t)}\left\{\int_0^t \frac{2}{u}e^{-2E_1(u)}\int_u^\infty e^{-\Omega(v)}dv - 2e^{-u}du\right\}dt, \\ U_{2,1}(s) = \int_0^s e^{-\Omega(t)}\left\{\int_0^t \frac{2}{u}e^{-2E_1(u)}du\right\}dt. \end{cases}$$

$$(4.25)$$

By using the expansion of $\exp(-\Omega(t))$ and $E_1(s)$ and integrating we get the expansions at 0:

$$\begin{cases} U_{1,1}(s) = -s + O(s^2), \\ U_{2,1}(s) = s + O(s^2). \end{cases}$$

Now we write

$$U(s,h) = a_1(h)U_1(s,h) + a_2(h)U_2(s,h)$$
$$= a_1(h)\{U_{1,0}(s) + hU_{1,1}(s) + \dots\}$$
$$\quad + a_2(h)\{U_{2,0}(s) + hU_{2,1}(s) + \dots\}$$

with a_1 and a_2 depending on h. In order to find the right values for a_1 and a_2, we have to use relation (4.21). Without loss of generality we may assume that $a_2(h) = 1$. Then by using (4.21) we get

$$0 = \lim_{h\to 0}\frac{U'(s,h)}{U(s,h)} = \lim_{h\to 0}\frac{hU_{2,1}'(s) + \dots + a_1(h)\left[U_{1,0}'(s) + \dots\right]}{1 + hU_{2,1}(s) + \dots + a_1(h)\left[U_{1,0} + \dots\right]},$$

from which one gets $\lim_{h\to 0}a_1(h) = 0$ and so one gets $\lim_{h\to 0}U(s,h) = 1$.

An integration by parts shows that

$$\int_0^s e^{-2E_1(t)}\frac{dt}{t} = \log s\, e^{-2E_1(s)} - 2\int_0^s \frac{\log t}{t}e^{-t}e^{-2E_1(t)}dt.$$

From which it follows that

$$\int_0^s e^{-2E_1(t)} \frac{dt}{t} = \log \ s - 2I + O\left(e^{-s}\right) \quad \text{for} \ \ s \gg 1, \qquad (4.26)$$

where

$$I = \int_0^\infty \frac{\log \ t}{t} e^{-t} \ e^{-2E_1(t)} \ dt.$$

Hence, from (4.25) and (4.26),

$$U'_{2,1}(s) = e^{-2s}(2 \ \log \ s - 4I) + O\left(e^{-3s}\right) \quad \text{for} \ s \gg 1.$$

We also have

$$U'_{1,0} = -e^{2\gamma}e^{-2s}(1 + O\left(e^{-s}\right)) \quad \text{for} \ s \gg 1.$$

So, we get for $s \gg 1$ and $hs \ll 1$ the estimates:

$$U'(s,h) = h \left\{ e^{-2s}(2 \ \log \ s - 4I) + O\left(e^{-3s}\right) + O(h) \right\} \\ + a_1(h) \left\{ e^{2\gamma}e^{-2s}(1 + O\left(e^{-s}\right)) + O(h) \right\}.$$

The only way for this estimate to be compatible with (4.21) is to have

$$a_1(h) = e^{-2\gamma}2h[\gamma - 1 + \log(2h) + 2I] + o(h) \text{ for } h \to 0.$$

So, we get the expansion of $sU(s,h)$ around $s = 0$ to be

$$sU(s,h) = s + hsU_{2,0}(s) + \cdots + a_1(h) \left[sU_{1,0}(s) + hsU_{1,1}(s) + O(h^2) \right], \\ = s + hO\left(s^2\right) + \cdots + a_1(h) \left[1 + O(s^2) + hO(s^2) + O(h^2) \right].$$

We can apply the implicit function theorem [Avez (1986)] with two variables and get an expansion of the root s_0 of $sU(s,h)$ in terms of $a_1(h)$ and h. By doing this expansion one gets that the zero of $sU(s,h)$ is

$$s_0 = -a_1(h) + O\left(h^2(\log(h))^2\right).$$

Since $a(h) = -s_0$ the result follows. □

The second order term of the expansion is obtained in [Morrison (1987)]. It would be interesting to compute to any order the expansion of $a(h)$.

4.5 Maximum of gaps

Let $Y(t)$ be the random variable which represents the maximum of length of gaps generated by random packing of unit intervals into $[0, t]$. We have

$$Pr(Y(x+1) \le h) = \frac{1}{x} \int_0^x Pr(Y(y) \le h)Pr(Y(x-y) \le h)dy$$

with

$$Pr(Y(x) \le h) = \begin{cases} 1 \text{ for } 0 \le x < h, \\ 0 \text{ for } h \le x < 1, \\ 1 \text{ for } x = 1. \end{cases}$$

Thus $Pr(Y(t) \le h)$ satisfies the same equation as $Pr(L(t) \ge h)$ and differ only from the initial condition. We indicate here some basic properties of this problem which awaits further study.

Let us define $f(t) = Pr(Y(t) \le h)$ and the Laplace transform

$$g(t) = \int_0^\infty e^{-sx} f(x+1)dx.$$

The function g satisfies the differential equation

$$\left(\frac{1 - e^{-sh}}{s} + e^{-s}g(s) \right)^2 = -\frac{d}{ds}g(s), \tag{4.27}$$

which can be proved by the same method as for the minimum of gap. Also we get that there exists a constant $b(h) > 0$ such that

$$\lim_{x \to \infty} \frac{1}{x} \int_0^x e^{b(h)(u+1)} Pr(Y(u) \le h)du = 1$$

by using Equation (4.27) and proving that its solution blows up at a certain point $b(h)$. Going further and obtaining estimates of $b(h)$ is an interesting problem.

4.6   Appendix: Renewal equations

We expose here without proof some results from Renewal Theory. This beautiful subject is explained in detail in [Feller (1968)] for the discrete case and in [Feller (1971)] for the continuous case. We expose here the continuous theory without giving any proof but trying to give the ideas involved.

Recall that if X is a random variable, which we assume to have a continuous probability density f, then we denote by

$$f^{n\star} = f \star f \star \cdots \star f$$

the probability density of the sum X_n of n independent copy of X. Here $\star$ is the convolution product of Definition 3.1. Note that $f^{0\star}$ can be interpreted as the probability measure δ_0 concentrated at 0. For f a probability density and z a real function on $\mathbb{R}_+$, the renewal equation is the following equation in Z:

$$z + Z \star f = Z. \tag{4.28}$$

The solution of such an equation ought to be

$$Z = z + z \star f + z \star f \star f + \cdots + z \star f^{n\star} + \cdots = \sum_{k=0}^{\infty} z \star f^{k\star}$$

as one can check by substitution in (4.28). But making sense of the series is not simple; a detailed exposition of the history and technical difficulties is available from [Feller (1968, 1971)].

We first define the iterated sum

$$Z_n = \sum_{k=0}^{n} z \star f^{k\star}.$$

If X is a positive random variable which is not identically 0 then it is easy to see that for every finite interval $[a, b]$ we have

$$\lim_{k \to \infty} \int_a^b f^{k\star}(t) dt = 0. \tag{4.29}$$

The reason is that $X > 0$ over a set of non-zero measure and this suffices to make the sum X_n of n independent copies of X converge to ∞ over a set of measure 1.

By the same kind of argument, one can give sense to the following limit:

$$U = \lim_{n \to \infty} U_n.$$

The following expansion

$$z + Z_n \star f - Z_n = z \star f^{(n+1)\star}$$

shows that one can consider Z_n to be an approximate solution of Equation (4.28). By using the limit (4.29) one can prove the convergence of Z_n towards a solution of the equation.

Let us introduce now the probability density of the exponential probability distribution. For arbitrary but fixed $\alpha > 0$ put

$$f(x) = \alpha e^{-\alpha x}, \quad F(x) = 1 - e^{-\alpha x}, \text{ for } x \geq 0 \tag{4.30}$$

and $F(x) = f(x) = 0$ for $x < 0$. Then f is an exponential density, F its distribution function. An easy calculation shows that the expectation μ equals α^{-1}, the variance α^{-2}.

If $T_1, \cdots, T_n$ are mutually independent random variables with the above exponential density 4.30, then the sum $T_1 + \cdots + T_n$ has a density g_n and distribution function G_n given by

$$
\begin{cases}
g_n(x) = \alpha \frac{(\alpha x)^{n-1}}{(n-1)!} e^{-\alpha x} & \text{for } x > 0, \\
G_n(x) = 1 - e^{-\alpha x} \left(1 + \frac{\alpha x}{1!} + \cdots + \frac{(\alpha x)^{n-1}}{(n-1)!} \right) & \text{for } x > 0.
\end{cases}
$$

We see that

$$
\sum_{n=1}^{\infty} f^{n\star}(x) = \sum_{n=1}^{\infty} g_n(x) = \sum_{n=1}^{\infty} \alpha \frac{(\alpha x)^{n-1}}{(n-1)!} e^{-\alpha x} = \alpha = \frac{1}{\mu},
$$

which shows that the infinite sum is constant.

Thus g_n is the probability density of n bus arriving at a time t. Thus in the case of exponential the sum

$$
\sum_{n=1}^{\infty} f^{n\star}(x)
$$

is interpreted as the density of bus at a time x, that is disregarding the order in which they are taken. That it is constant is an expected result and Renewal Theory seeks to generalize this to general distributions.

Theorem 4.9. *If f is a probability density and z is absolutely integrable then Equation (4.28) has a unique solution Z which satisfies the limit*

$$
\lim_{x \to \infty} Z(x) = \frac{1}{\mu} \int_0^{\infty} z(x) dx
$$

with

$$
\mu = \int_0^{\infty} t f(t) dt
$$

if the integral converge and $\mu = \infty$ otherwise.

We refer to [Feller (1971)] for the proof of this theorem and for examples illustrating it.

However, in our application we need to slightly generalize the theorem by considering non-negative integrable functions f with $\int_0^{\infty} f(t) dt < 1$.

Theorem 4.10. *[Feller (1971)] Let f be an integrable non-negative function on $\mathbb{R}_+$ with $F_\infty = \int_0^{\infty} f(t) dt < 1$. Let f be a real function on $\mathbb{R}_+$ which admits a limit at ∞. Then the solution $Z(x)$ of the equation*

$$
Z(x) = z(x) + \int_0^x Z(x-t) f(t) dt
$$

satisfies

$$
\lim_{t \to \infty} Z(t) = \frac{1}{1 - F_\infty} \lim_{t \to \infty} z(t).
$$

So, if z converges to a value $z(\infty)$ then Z converges to $Z(\infty) = \frac{1}{1-F_\infty}z(\infty)$. In the case where z has limit 0 at ∞ it is possible to estimate the rate of decay of Z to 0:

Theorem 4.11. *Let f be an integrable function on $\mathbb{R}_+$ with $F_\infty = \int_0^\infty f(t)dt < 1$. Let z be a real function on $\mathbb{R}_+$ with $\lim_{x\to\infty} z(x) = 0$. Then the solution $Z(x)$ of the equation*

$$Z(x) = z(x) + \int_0^x Z(x-t)f(t)dt$$

satisfies to the limit

$$\lim_{x\to\infty} Z(x)e^{kx} = \frac{\int_0^\infty e^{kx}z(x)dx}{\mu^\sharp}$$

with k the unique real solution of

$$\int_0^\infty e^{kt}f(t)dt = 1$$

and

$$\mu^\sharp = \int_0^\infty te^{kt}f(t)dt.$$

Proof. Let us define

$$Z^\sharp(x) = e^{kx}Z(x), \quad z^\sharp(x) = e^{kx}z(x) \text{ and } f^\sharp(t) = e^{kt}f(t).$$

We obtain the equation

$$Z^\sharp(x) = z^\sharp(x) + \int_0^x Z^\sharp(x-t)f^\sharp(t)dt.$$

So, by the renewal theorem 4.9 we have

$$\lim_{x\to\infty} Z^\sharp(x) = \frac{1}{\mu^\sharp}\int_0^\infty e^{kx}z(x)dx$$

with

$$\mu^\sharp = \int_0^\infty e^{kt}tf(t)dt.$$

So, finally we get an estimate $Z(t) \simeq ae^{-kt}$ as $t \to \infty$ for some $a > 0$. $\quad\square$

If $\mu^\sharp = \infty$ then the theorem says that $\lim_{x\to\infty} Z(x)e^{kx} = 0$. In case $\mu^\sharp < \infty$ it gives the equivalent

$$Z(x) \simeq \frac{1}{\mu^\sharp}e^{-kx}$$

of the solution Z.

Chapter 5

Integral equation method for the 1-dimensional random packing

We give another proof of the existence of the limit density for the 1-dimensional random packing problem. Our proof is based on [Dvoretzky and Robbins (1964)] and is completely independent of Rényi's method. Our presentation follows very closely theirs since we were not able to improve it. This method allows one to prove a very fast convergence of the quotient $\frac{M(x)}{x}$ towards Rényi's constant with a simple use of analytical methods. Thus this method is also used for proving a central limit theorem and in Chapter 8 to prove the convergence of another sequential random packing process.

Let us remind the reader some notation from Chapter 3. We denote by N_x the random variable of the number of cubes put in Rényi's random cube packing procedure and $M(x) = E(N_x)$. If one defines the function

$$f(x) = M(x) + 1$$

then we see that f satisfies the somewhat simpler equation

$$f(x+1) = \frac{2}{x} \int_0^x f(t)dt \text{ with } x > 0. \tag{5.1}$$

Together with the initial conditions

$$f(x) = \begin{cases} 1 & \text{if } 0 \le x < 1, \\ 2 & \text{if } x = 1. \end{cases}$$

We will prove that $L(x) = C_R x + C_R - 1$ approximates $M(x)$ very well.

5.1 Estimating $M(x) - C_R x$

We first state and prove the general results on the integral equations that we will use later:

Theorem 5.1. *[Dvoretzky and Robbins (1964)] Let $f(x)$ be defined for $x \geq 0$ and satisfy*

$$f(x+1) = \frac{2}{x}\int_0^x f(t)dt + p(x+1) \text{ for } x > 0 \qquad (5.2)$$

where $p(x)$ is continuous for $x > 1$ and is such that, setting

$$p_x = \sup_{x \leq t \leq x+1} |p(t)| \quad \text{for } x > 1 \qquad (5.3)$$

we have

$$\sum_{i=2}^{\infty} \frac{p_i}{i} < \infty. \qquad (5.4)$$

Then there exists a constant λ such that, setting

$$R_j = \frac{2j+1}{j}p_{j+1} + \frac{2(j+1)(j+3)}{j}\sum_{i=j+2}^{\infty}\frac{p_j}{i+1} \text{ for } j = 1,2,\ldots \qquad (5.5)$$

we have for $n = 1, 2, \ldots$

$$\sup_{n+1 \leq x \leq n+2}|f(x) - \lambda x - \lambda| \leq \frac{2^n}{n!}\sup_{1 \leq x \leq 2}|f(x) - \lambda x - \lambda| \atop + \frac{2^a}{n!}\sum_{j=1}^{n}\frac{j!}{2^j}R_j. \qquad (5.6)$$

Proof.   From (5.2) we have for positive x and y,

$$f(y+1) = \frac{2}{y}\int_0^x f(t)dt + \frac{2}{y}\int_x^y f(t)dt + p(y+1)$$
$$= \frac{1}{y}[xf(x+1) - xp(x+1)] + \frac{2}{y}\int_x^y f(t)dt + p(y+1)$$

or

$$f(y+1) = \frac{x}{y}f(x+1) + \frac{2}{y}\int_x^y f(t)dt + p(y+1) - \frac{x}{y}p(x+1). \qquad (5.7)$$

Define

$$I_x = \inf_{x \leq t \leq x+1}\frac{f(t)}{t+1}, \quad S_x = \sup_{x \leq t \leq x+1}\frac{f(t)}{t+1} \quad \text{for } x \geq 0. \qquad (5.8)$$

Notice that $f(x) = x + 1$ satisfies (5.2) with $p = 0$, and hence that

$$y + 2 = \frac{2}{y}(x+2) + \frac{2}{y}\int_x^y(t+1)dt. \qquad (5.9)$$

Subtracting (5.9) multiplied by I_x from (5.7) we have

$$f(y+1) - I_x \cdot (y+2) = \frac{x}{y}[f(x+1) - I_x \cdot (x+2)] \atop + \frac{2}{y}\int_x^y[f(t) - I_x \cdot (t+1)]dt \atop + p(y+1) - \frac{x}{y}p(x+1).$$

Hence for $x \leq y \leq x + 1$, in view of (5.8) and (5.3),

$$f(y+1) - I_x \cdot (y+2) \geq 0 + 0 - p_{x+1} - p_{x+1} = -2p_{x+1}.$$

It follows that

$$I_{x+1} \geq I_x - \frac{2p_{x+1}}{x+2} \text{ for } x > 0. \tag{5.10}$$

Applying (5.10) successively with x replaced by $x+1, x+2, \ldots$ we obtain

$$I_y \geq I_x - \Delta_x, \text{ for } y \geq x > 0 \tag{5.11}$$

where by definition

$$\Delta_x = 2 \sum_{i=1}^{\infty} \frac{p_{x+i}}{x+i+1} \text{ for } x > 0.$$

In exactly the same manner we obtain the inequality

$$S_y \leq S_x + \Delta_x \text{ for } y \geq x > 0.$$

From (5.11) we have

$$\underline{\lim}_{y \to \infty} I_y \geq I_x - \Delta_x$$

where $\underline{\lim}_{y \to \infty} I_y = \lim_{x \to \infty} \inf_{y \geq x} I_y$ is the inferior limit of I_y.
Since $\Delta_x = o(1)$ by (5.4) it follows that

$$\underline{\lim}_{y \to \infty} I_y \geq \overline{\lim}_{x \to \infty} I_x.$$

From this and (5.11) with $x = 1$ we find that

$$I_\infty = \lim_{x \to \infty} I_x$$

exists, and $I_\infty > -\infty$. Similarly

$$S_\infty = \lim_{x \to \infty} S_x$$

exists, and $S_\infty < \infty$.
Since $I_x \leq S_x$ it follows that

$$-\infty < I_\infty \leq S_\infty < \infty. \tag{5.12}$$

From (5.7) we have for $x, y > 0$

$$\begin{aligned} f(y+1) - f(x+1) &= \tfrac{x-y}{y} f(x+1) + \tfrac{2}{y} \int_x^y f(t) dt \\ &+ p(y+1) - \tfrac{x}{y} p(x+1). \end{aligned} \tag{5.13}$$

By (5.8) and (5.12), $f(x) = O(x)$, and hence by (5.13)

$$\sup_{x \leq y \leq x+1} |f(y+1) - f(x+1)| = O(1) + 2p_x.$$

But this implies by (5.4) that

$$S_x - I_x = o(1)$$

and therefore that

$$I_\infty = S_\infty \neq \pm\infty. \tag{5.14}$$

We now define λ as the common value in (5.14).

$$\lambda = \lim_{x\to\infty} I_x = \lim_{x\to\infty} S_x = \lim_{x\to\infty} \frac{f(x)}{x+1}. \tag{5.15}$$

By (5.11) and (5.15),

$$I_x - \Delta_x \leq \lambda \leq S_x + \Delta_x \text{ for } x > 0. \tag{5.16}$$

Next we observe that for every $x > 1$ there exists a number x' satisfying

$$x \leq x' \leq x + 1 \text{ and } \left|\frac{f(x')}{x'+1} - \lambda\right| \leq \Delta_x. \tag{5.17}$$

Indeed, since by (5.2) $f(x)$ is continuous for $x > 1$ the non-existence of such an x' would imply that either

$$I_x > \lambda + \Delta_x \text{ or } S_x < \lambda - \Delta_x,$$

contradicting (5.16). We denote by x_n a value x' satisfying (5.17) for $x = n$; thus for $n = 2, 3, \ldots$

$$|f(x_n) - \lambda(x_n + 1)| \leq (n+2)\Delta_n \text{ for } n \leq x_n \leq n + 1. \tag{5.18}$$

Now set

$$f^*(x) = f(x) - \lambda(x + 1)$$

then f^* again satisfies (5.2), and applying (5.7) with $n \leq y \leq n + 1$ and $x = x_{n+1} - 1$ we obtain from (5.18) for $n = 1, 2, \ldots$

$$|f^*(y + 1)| \leq \frac{n+1}{n}(n + 3)\Delta_{n+1} + \frac{2}{n}\sup_{n \leq t \leq n+1} |f^*(t)| + p_{n+1} + \frac{n+1}{n}p_{n+1}. \tag{5.19}$$

Putting

$$T_x = \sup_{x \leq t \leq x+1} |f^*(t)| \text{ for } x > 0, \tag{5.20}$$

we obtain from (5.19) for $n = 1, 2, \ldots$:

$$T_{n+1} \leq \frac{2}{n}T_n + \frac{2n+1}{n}p_{n+1} + \frac{(n+1)(n+3)}{n}\Delta_{n+1} = \frac{2}{n}T_n + R_n$$

where R_n is defined by (5.5). Successive applications of this inequality for $n = 1, 2, 3, \ldots$ yield the inequality

$$T_{n+1} \le \frac{2^n}{n!} T_1 + \frac{2^n}{n!} \left[\frac{1!}{2} R_1 + \frac{2!}{2^2} R_2 + \ldots + \frac{n!}{2^n} R_n \right].$$

In view of (5.20) this is precisely (5.6). $\qquad\square$

Corollary 5.1. *[Dvoretzky and Robbins (1964)] If $a > 2e$ and $f(x)$ satisfies* (5.2) *with*

$$p(x) = O\left(\left(\frac{a}{x}\right)^{x+\beta} \right), \tag{5.21}$$

then

$$f(x) = \lambda x + \lambda + O\left(\left(\frac{a}{x}\right)^{x+\beta-1} \right). \tag{5.22}$$

Proof.   If (5.21) holds, then by (5.5)

$$R_j = O\left(\left(\frac{a}{j}\right)^{j+\beta+1} \right)$$

and hence (5.6), since $\alpha > 2e$,

$$\frac{2^n}{n!} \sum_{j=1}^{n} \frac{j!}{2^j} R_j = O\left(\left(\frac{\alpha}{n}\right)^{j+\beta+1} \right).$$

Thus by (5.6)

$$\begin{aligned} \sup_{n+1 \le x \le n+2} | f(x) - \lambda x - \lambda | &= O\left(\left(\frac{2e}{n}\right)^{n+\frac{1}{2}} \right) + O\left(\left(\frac{\alpha}{n}\right)^{n+\beta+1} \right) \\ &= O\left(\left(\frac{\alpha}{n}\right)^{n+\beta+1} \right) \end{aligned}$$

from which (5.22) follows. $\qquad\square$

Theorem 5.2. *[Dvoretzky and Robbins (1964)] Let $g(x)$ be defined for $x \ge 0$ and satisfy*

$$g(x+1) = \frac{2}{x} \int_0^x g(t)dt + O\left(x^\gamma\right) \quad \text{for } x > 0$$

with $\gamma > 1$. Then

$$g(x) = O\left(x^\gamma\right). \tag{5.23}$$

Proof. We have

$$g(x+1) = \frac{2}{x} \int_0^x g(t)dt + \eta(x) \text{ with } x > 0$$

where

$$\eta(x) = O\left(x^\gamma\right) \text{ for } \gamma > 1.$$

Choose $x_0 > 1$ and $H > 0$ such that

$$|\eta(x)| \le Hx^\gamma \quad \text{for } x \ge x_0 - 1,$$
$$\text{and } \int_0^{x_0} |g(t)| \, dt \le \frac{H}{\gamma - 1}(x_0 - 1)^{\gamma + 1} = \frac{\gamma + 1}{\gamma - 1} H \int_0^{x_0 - 1} t^\gamma dt. \tag{5.24}$$

Then for $x_0 - 1 \le x \le x_0$ we have

$$\begin{aligned}
|g(x+1)| &\le \frac{2}{x} \int_0^{x_0} |g(t)| \, dt + Hx^\gamma \\
&\le \frac{2H}{x(\gamma - 1)}(x_0 - 1)^{\gamma + 1} + Hx^\gamma \\
&\le \frac{2Hx^\gamma}{\gamma - 1} + Hx^\gamma = \frac{\gamma + 1}{\gamma - 1} Hx^\gamma.
\end{aligned} \tag{5.25}$$

Hence

$$\begin{aligned}
\int_0^{x_0 + 1} |g(t)| dt &= \int_0^{x_0} |g(t)| dt + \int_{x_0}^{x_0 + 1} |g(t)| dt \\
&\le \frac{\gamma + 1}{\gamma - 1} H \int_t^{x_0} (t - 1)^\gamma dt + \int_{x_0}^{x_0 + 1} \frac{\gamma + 1}{\gamma - 1} H(t - 1)^\gamma dt \\
&= \frac{\gamma + 1}{\gamma - 1} H \int_0^{x_0} t^\gamma dt,
\end{aligned}$$

so that (5.24) holds with x_0 replaced by $x_0 + 1$. Hence by (5.25), for $x_0 \le x \le x_0 + 1$ we have

$$|g(x+1)| \le \frac{\gamma + 1}{\gamma - 1} Hx^\gamma. \tag{5.26}$$

By induction, (5.26) holds for all $x \ge x_0 - 1$, which proves (5.23). $\square$

Corollary 5.2. *[Dvoretzky and Robbins (1964)] Let $g(x)$ be defined for $x \ge 0$ and satisfy*

$$g(x+1) = \frac{2}{x} \int_0^x g(t)dt + Ax^\beta + O\left(x^\gamma\right)$$

with $\beta > \gamma > 1$. Then

$$g(x) = \frac{\beta + 1}{\beta - 1} Ax^\beta + O\left(x^{\max(\beta - 1, \gamma)}\right).$$

Proof. Set

$$g^*(x) = \frac{\beta + 1}{\beta - 1} Ax^\beta.$$

Then

$$\begin{aligned}
g^*(x+1) &= \frac{\beta + 1}{\beta - 1} A(x + 1)^\beta = \frac{\beta + 1}{\beta - 1} Ax^\beta + O\left(x^{\beta - 1}\right) \\
&= \frac{2}{x} \int_0^x g^*(t)dt + Ax^\beta + O\left(x^{\beta - 1}\right).
\end{aligned}$$

Hence, setting

$$\bar{g}(x) = g(x) - g^*(x)$$

we have for $x > 0$,

$$\bar{g}(x+1) = g(x+1) - g^*(x+1) = \frac{2}{x}\int_0^x \bar{g}(t)dt + O\left(x^{\max(\beta-1,\gamma)}\right).$$

Hence by Theorem 5.2 we have

$$\bar{g}(x) = O\left(x^{\max(\beta-1,\gamma)}\right)$$

which is required. □

Theorem 5.3. *The expectation $M(x)$ of Rényi's random cube packing process satisfies the relation*

$$\sup_{n+1\leq x\leq n+2} |M(x) - (C_R x + (C_R - 1))| < \frac{2^n}{n!}$$

for $n = 0, 1, \ldots$.

Proof. Since $f(x) = M(x) + 1$ satisfies by (5.1), Equation (5.2) with $p = 0$, we have by Theorem 5.1 another proof that

$$\lim_{x\to\infty} \frac{M(x)}{x}$$

exists, and by (5.16) for every $x > 0$,

$$\inf_{x\leq t\leq x+1} \frac{M(t)+1}{t+1} = I_x \leq C_R \leq S_x = \sup_{x\leq t\leq x+1} \frac{M(t)+1}{t+1}. \qquad (5.27)$$

Taking $x = 2$ we obtain easily from (3.3) that

$$0.666\cdots = \frac{2}{3} \leq C_R \leq 3 - \sqrt{5} = 0.75\ldots$$

and we can get tighter bound for larger values of x.

Since $M(x) = 1$ for $1 \leq x \leq 2$, even the crude approximation $\frac{1}{2} < C_R < 1$ yields

$$\sup_{1\leq x\leq 2} |M(x) + 1 - C_R x - C_R| = \max_{1\leq x\leq 2} |2 - C_R x - C_R|$$
$$= \max(|2 - 2C_R|, |2 - 3C_R|) \leq 1.$$

Hence the conclusion follows from Theorem 5.1 with $p = 0$. □

By using Equation (5.27) the value of C_R is estimated in [Blaisdell and Solomon (1970)] up to 15 decimals by computing $M(t)$ for $14 \leq t \leq 15$.

By Stirling's formula it follows that

$$M(x) = C_R x + C_R - 1 + O\left(\left(\frac{2e}{x}\right)^{x-3/2}\right). \qquad (5.28)$$

This is actually a more precise result than the one obtained by Rényi's method in Section 3.4, Theorem 3.4.

5.2 The variance and the central limit theorem

We denote by $V(X) = E\left(X^2\right) - E(X)^2$ the variance of a random variable, $\sigma(X) = \sqrt{V(X)}$ its standard deviation and we have Markov's inequality

$$E\left(|X - E(X)| \geq a\right) \leq \frac{1}{a}\sigma(X).$$

We first prove a general theorem on the variance $V(N_x)$ of N_x:

Theorem 5.4. *[Dvoretzky and Robbins (1964)] The variance satisfies the integral inequality estimate*

$$V(N_{x+1}) \geq \frac{1}{x}\int_0^x V(N_t) + V(N_{x-t})dt = \frac{2}{x}\int_0^x V(N_t)dt. \qquad (5.29)$$

Proof. We consider the parking of cars of length 1 in $[0, x+1]$. For $1 \leq x$ we condition on the first interval $[Y, Y + 1]$ since the number of intervals N_t in $[0, t]$, and $N_{(x+1)-(t+1)}$ in $[t + 1, x + 1]$ are mutually independent if $Y = t$. The variance for $0 \leq t \leq x$ is given by

$$\begin{aligned}
&V(N_{x+1} \mid Y = t)\\
&= \sum_{n_1, n_2} (n_1 + n_2 - M(t) - M(x - t))^2 Pr(N_t = n_1)Pr(N_{x-t} = n_2)\\
&= \sum_{n_1}(n_1 - M(t))^2 Pr(N_t = n_1) + \sum_{n_2}(n_2 - M(x - t))^2 Pr(N_{x-t} = n_2)\\
&\quad +2\sum_{n_1, n_2}(n_1 - M(t))(n_2 - M(x - t))Pr(N_t = n_1)Pr(N_{x-t} = n_2)\\
&= V(N_t) + V(N_{x-t}), \qquad (5.30)
\end{aligned}$$

since

$$2\sum_{n_1, n_2}(n_1 - M(t))(n_2 - M(x - t))Pr(N_t = n_1)Pr(N_{x-t} = n_2) = 0.$$

Let $Pr(Y = t)$ be the probability density of the random variable Y at t.

We have

$$
\begin{aligned}
V(N_{x+1}) &= \int_0^x \Big\{ \textstyle\sum_{n_1,n_2} (n_1 + n_2 + 1 - M(x+1))^2 Pr(N_t = n_1) \\
&\quad Pr(N_{x-t} = n_2) \Big\} Pr(Y = t)dt \\
&= \int_0^x \Big\{ \textstyle\sum_{n_1,n_2} (n_1 + n_2 + 1 - M(t) - M(x-t) \\
&\quad + M(t) + M(x-t) - M(x+1))^2 \\
&\quad \times Pr(N_t = n_1)Pr(N_{x-t} = n_2) \Big\} Pr(Y = t)dt \\
&= \int_0^x [\textstyle\sum_{n_1,n_2} \Big((n_1 + n_2 - M(t) - M(x-t))^2 \\
&\quad + 2(n_1 + n_2 - M(t) - M(x-t)) \\
&\quad (M(t) + M(x-t) + 1 - M(x+1)) \\
&\quad + (M(t) + M(x-t) + 1 - M(x+1))^2 \Big) \\
&\quad \times Pr(N_t = n_1)Pr(N_{x-t} = n_2)]Pr(Y = t)dt.
\end{aligned}
$$

From (5.30)

$$
V(N_t) + V(N_{x-t}) = \textstyle\sum_{n_1,n_2} (n_1 + n_2 - M(t) - M(x-t))^2 \\
Pr(N_t = n_1)Pr(N_{x-t} = n_2)
$$

and

$$
0 = \textstyle\sum_{n_1,n_2} 2(n_1 + n_2 - M(t) - M(x-t))(M(t) + M(x-t) + \\
1 - M(x+1)) \times Pr(N_t = n_1)Pr(N_{x-t} = n_2).
$$

We have

$$
V(N_{x+1}) = \int_0^x \big(V(N_t) + V(N_{x-t}) \\
+ (M(t) + M(x-t) + 1 - M(x+1))^2 \big) Pr(Y = t)dt, \tag{5.31}
$$

which gives the required relation. □

Theorem 5.5. *[Rényi (1958)] For $M(x) = E(N_x)$ and the second moment $M_2(x) = E(N_x^2)$, we have*

$$
M_2(x+1) = 1 + \frac{2}{x}\int_0^x M_2(t)dt + \frac{4}{x}\int_0^x M(t)dt + \frac{2}{x}\int_0^x M(t)M(x-t)dt. \tag{5.32}
$$

Proof. Conditioning on the first interval is $[t, t+1]$, we have two remaining intervals $[0, t]$ and $[t+1, x+1]$ of length t, respectively $x-t$. The random variables N_{x+1}, N_t and N_{x-t} of the number of intervals in $[0, x+1]$, $[0, t]$ and $[t+1, x+1]$, where N_t and N_{x-t} are mutually independent. We have

$$
M(x+1) = \frac{1}{x}\int_0^x E(N_{x+1} \mid t)dt = \frac{1}{x}\int_0^x E(1 + N_t + N_{x-t})dt,
$$

which gives (3.2). We have for the second moment,

$$
\begin{aligned}
M_2(x+1) &= E\left(N_{x+1}^2\right) \\
&= \tfrac{1}{x}\int_0^x E\left(N_{x+1}^2 \mid t\right) dt \\
&= \tfrac{1}{x}\int_0^x E\left((1 + N_t + N_{x-t})^2\right) \mid t\right) dt \\
&= \tfrac{1}{x}\int_0^x E\left(1 + 2N_t + 2N_{x-t} + 2N_t N_{x-t}\right. \\
&\quad \left. + N_t^2 + N_{x-t}^2 \mid t\right) dt \\
&= \tfrac{1}{x}\left\{\int_0^x 1 + 2M(t) + 2M(x-t) + M_2(t) + M_2(x-t)dt\right. \\
&\quad \left. + 2\int_0^x M(t)M(x-t)dt\right\} \\
&= 1 + \tfrac{1}{x}\int_0^x 4M(t) + 2M_2(t) + 2M(t)M(x-t)dt.
\end{aligned}
$$

This is the required result.   □

By using Equation (5.32) it is possible to compute $E\left(N_x^2\right)$ and thus $V\left(N_x\right)$. The plot is given in Figure 5.1.

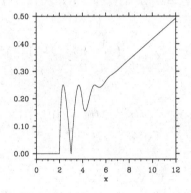

Fig. 5.1   The variance $V(N_x)$ of Rényi's sequential random cube packing process.

Let

$$
L(x) = C_R x + C_R - 1,
$$

where C_R is Rényi's packing constant of Chapter 3, and define for $k = 1, 2, \ldots$

$$
\phi_k(x) = E\left((N_x - L(x))^k\right). \tag{5.33}
$$

Since

$$
L(x+1) = L(t) + L(x-t) + 1,
$$

we have

$$
E\left[(N_{x+1} - L(x+1))^k \mid t\right] = E\left[\{(N_t - L(t))(N_{x-t} - L(x-t))\}^k\right]
$$

and by integration we find that

$$\phi_k(x+1) = \frac{1}{x}\sum_{i=0}^{k}\binom{k}{i}\int_0^x \phi_i\phi_{k-i}(x-t)dt \text{ for } x > 0. \tag{5.34}$$

We first estimate the variance:

Theorem 5.6. *[Dvoretzky and Robbins (1964)] There exists a constant $\lambda_2 > 0$ such that the variance $V(N_x)$ of N_x satisfies the relation*

$$V(N_x) = \lambda_2 x + \lambda_2 + O\left(\left(\frac{4e}{x}\right)^{x-4}\right). \tag{5.35}$$

Proof. By (5.34) with $k = 2$ we have for $x > 0$

$$\phi_2(x+1) = \frac{2}{x}\int_0^x \phi_2(t)dt + \frac{2}{x}\int_0^x \phi_1(t)\phi_1(x-t)dt.$$

But $\phi_1(t)$ is estimated by (5.28), and therefore

$$\sup_{0<t<x} |\phi_1(t)\phi_1(x-t)| = O\left(\left(\frac{4e}{x}\right)^{x-3}\right). \tag{5.36}$$

But $f(x) = \phi_2(x)$ satisfies (5.2) with p estimated by (5.36). $\phi_2(x)$ satisfies (5.35) by the Corollary to Theorem 5.1, and

$$V(N_x) - \phi_2(x) = -(\phi_1(x))^2,$$

which by (5.28), is absorbed into the error term. It remains to show that $\lambda_2 > 0$. This may be done numerically from estimates obtained in the course of the proof of Theorem 5.1, but it is much simpler to deduce it as follows. Since $V(N_x) > 0$ for $2 < x < 3$ it follows from Theorem 5.4 that $V(N_x) > \frac{\delta}{x}$ for some $\delta > 0$. But this contradicts (5.35) unless $\lambda_2 > 0$. $\quad\square$

In [Blaisdell and Solomon (1970)] λ_2 is estimated to be about 0.038155. We now prove a result on the central moments of N_x.

Theorem 5.7. *[Dvoretzky and Robbins (1964)] For every $k = 1, 2, \ldots$ and $\epsilon > 0$,*

$$E\left((N_x - M(x))^k\right) = c_k x^{\lfloor \frac{k}{2} \rfloor} + O\left(x^{\lfloor \frac{k}{2} \rfloor - 1 + \epsilon}\right), \tag{5.37}$$

where the c_k are constants and

$$c_{2k} = \frac{(2k)!}{2^k k!}\lambda_2^k. \tag{5.38}$$

Proof. Since by (5.33)

$$N_x - M(x) = N_x - L(x) - \phi_1(x),$$

for $k = 1$ it follows from (5.28) that (5.37) is equivalent to

$$\phi_k(x) = c_k x^{\lfloor \frac{k}{2} \rfloor} + O\left(x^{\lfloor \frac{k}{2} \rfloor - 1 + \epsilon}\right). \tag{5.39}$$

By (5.28) and (5.35), (5.37) with (5.38) holds for $k = 1, 2$. By (5.34)

$$\phi_3(x+1) = \frac{2}{x} \int_0^x \phi_3(t)dt + \frac{6}{x} \int_0^x \phi_1(t)\phi_2(x-t)dt,$$

and by (5.28) and (5.35) the second integral is $O\left(\left(\frac{C}{x}\right)^x\right)$ with a suitable C. Hence ϕ_3 satisfies (5.2) with p estimated as in (5.21). It follows from (5.22) that $\phi_3(x) = c_3 x + O(1)$ and thus (5.39) holds for $k \le 3$.

Now let $m > 3$ and assume that (5.39) holds for $k < m$. Then by (5.34),

$$\phi_m(x+1) = \frac{2}{x} \int_0^x \phi_m(t)dt + \frac{1}{x} \sum_{i=1}^{m-1} \binom{m}{i} \int_0^x \phi_i(t)\phi_{m-i}(x-t)dt. \tag{5.40}$$

By the induction assumption

$$\phi_i(t)\phi_{m-i}(x-t) = c_i c_{m-i} t^{\lfloor \frac{i}{2} \rfloor}(x-t)^{\lfloor \frac{m-i}{2} \rfloor} + O\left(x^{\lfloor \frac{i}{2} \rfloor + \lfloor \frac{m-i}{2} \rfloor - 1 + \epsilon}\right). \tag{5.41}$$

We have

$$\frac{1}{x} \int_0^x t^{\lfloor \frac{i}{2} \rfloor}(x-t)^{\lfloor \frac{m-i}{2} \rfloor} dt = \frac{(\lfloor \frac{i}{2} \rfloor)!(\lfloor \frac{m-i}{2} \rfloor)!}{(\lfloor \frac{i}{2} \rfloor + \lfloor \frac{m-i}{2} \rfloor + 1)!} x^{\lfloor \frac{i}{2} \rfloor + \lfloor \frac{m-i}{2} \rfloor}. \tag{5.42}$$

Since

$$\max_{1 \le i \le m-1} \left\lfloor \frac{i}{2} \right\rfloor + \left\lfloor \frac{m-i}{2} \right\rfloor = \left\lfloor \frac{m}{2} \right\rfloor \quad \text{for } m \ge 3$$

the sum of the right-hand side of (5.40) is

$$Cx^{\lfloor \frac{m}{2} \rfloor} + O\left(x^{\lfloor \frac{m}{2} \rfloor - 1 + \epsilon}\right) \tag{5.43}$$

with C some constant. Since $\lfloor \frac{m}{2} \rfloor \ge 2$ for $m > 3$, (5.39) for $k = m$ follows from (5.40) by the corollary of Theorem 5.2. Thus (5.39) holds for all $k = 1, 2, \ldots$. By (5.40), (5.41) and (5.42) the constant C in (5.43) for $m = 2k$ is

$$\sum_{j=1}^{k-1} \binom{2k}{2j} \frac{(k-j)!}{(k+1)!} c_{2j} c_{2k-2j}. \tag{5.44}$$

Assume that (5.38) holds for $c_2, \ldots, c_{2k-2}$. By (5.44) the coefficient of x^k in the equation

$$\phi_{2k}(x+1) = \frac{2}{x} \int_0^x \phi_{2k}(t)dt + Cx^k + O\left(x^{k-1+\epsilon}\right)$$

is

$$\frac{(k-1)(2k)!}{(k+1)!2^k}\lambda_2^k.$$

Thus by the Corollary of Theorem 5.2

$$\phi_{2k}(x) = \frac{k+1}{k-1}\frac{(k-1)(2k)!}{(k+1)!2^k}\lambda_2^k x^k + O\left(x^{k-1+\epsilon}\right)$$

and hence

$$c_{2k} = \frac{(2k)!}{k!2^k}\lambda_2^k$$

so that (5.38) holds for all $k = 1, 2, \cdots$.   □

Theorem 5.8. *[Dvoretzky and Robbins (1964)] The random variable*

$$Z_x = \frac{N_x - M(x)}{\sqrt{V(N_x)}}$$

is asymptotically normal of mean 0 and standard deviation 1 as $x \to \infty$.

Proof.   By Equations (5.37), (5.39) and (5.35) for $\epsilon = \frac{1}{2}$,

$$E\left(Z_x^k\right) = \frac{c_k x^{\lfloor\frac{k}{2}\rfloor} + o\left(x^{\lfloor\frac{k}{2}\rfloor}\right)}{(\lambda_2 x + o(x))^{k/2}}.$$

Hence

$$\lim_{x\to\infty} E\left(Z_x^k\right) = \begin{cases} \dfrac{k!}{2^{\frac{k}{2}}\left(\frac{k}{2}\right)!} & \text{if } k \text{ is even,} \\ 0 & \text{if } k \text{ is odd.} \end{cases}$$

Since these are the moments of the normal distribution of mean 0 and standard deviation 1, which is uniquely determined by its moments, the theorem follows from the moment convergence theorem.   □

Note that in [Dvoretzky and Robbins (1964)] two proofs of the above theorem are given. We make use of the speciality of the problem to give the one with moments computations. Another proof based on Lyapunov method which is a well-known standard method to prove the central limit theorem.

Chapter 6

Random sequential bisection and its associated binary tree

To any random sequential 1-dimensional packing process, it is possible to associate a corresponding random bisection tree. Here we analyze the obtained tree in the specific case of Kakutani's random sequential process. The Kakutani random splitting procedure [Kakutani (1975)] was originally proposed in the following way: consider an interval $[0, 1]$ and put a point x_1 at random in it. Then subdivide the interval of largest size by adding a point x_2 in it. The process is continued indefinitely, where at each step a point x_i is added at random in the middle of the interval of largest length. Kakutani asked whether the empirical distributions

$$F_n = \frac{1}{n} \sum_{k=1}^{n} \delta_{x_k}$$

converge to the uniform distribution on $[0, 1]$. *A priori* since the points are added where they are most needed, i.e. in the largest gaps, one would expect a very fast convergence. The stopped Kakutani's interval splitting is the interval splitting where one stops adding points when all interval lengths are lower than some number $t \leq 1$. Since all intervals are independent, it is apparent that the probability density of the stopped Kakutani's interval splitting is the same as the random sequential packing of intervals of length 0 in $[0, 1]$, where one stops inserting point when the gaps are smaller than, say, t. This remark allowed [Van Zwet (1978)] to prove the convergence of the empirical distribution of Kakutani's interval splitting to the uniform distribution. The stopping rule is discussed further in [Lootgieter (1977); Pyke (1980); Van Zwet (1978); Slud (1978)]. More results on random sequential bisection were obtained in [Devroye (1986)].

To study the asymptotic behavior of the lengths of subintervals in random sequential bisection, the associated binary tree is introduced. The number of internal or external nodes of the tree is asymptotically normal.

The levels of the lowest and the highest external nodes are bounded with probability one or with probability increasing to one as the number of nodes increases to infinity [Sibuya and Itoh (1987)].

The associated binary tree is closely related to random binary tree which arises in computer algorithms, such as binary search tree and quick sort.

6.1   Random sequential bisection

A binary tree in general is a finite set of nodes, which is partitioned, if not empty, into a triple; "root", "left subtree" and "right subtree", where the root is a subset of a single node and the subtrees feature is different from a usual tree which appears, for example, in cluster analysis. The notion has emerged from computer science techniques such as binary search tree and quick sort. See [Knuth (1975)], Vol. 1, Chapter 2, Sections 3 and 5 of this chapter. Other useful references on the notion of binary search tree are [Drmota (2009); Mahmoud (1992); Knuth (1973); Flajolet and Sedgewick (2009); Flajolet and Odlyzko (1982)].

Let a sequence $x_1, x_2, \cdots, x_n \in (0, x)$ of cars with length $l = 0$ be generated by the above random sequential packing in saturation. The level of an interval is sequentially defined. The interval $(0, x)$ with no parking car is at level 0. Consider the first t cars with $t \geq 1$. There are $t + 1$ gaps generated by the t cars. Let one of the $t + 1$ gaps, (α, β), $\alpha < \beta$, $\alpha, \beta \in \{0, x_1, x_2, \cdots, x_n, x\}$, at level d, be divided by the $t + 1$-th car $(\gamma = x_{t+1})$ into the left subinterval (α, γ) and the right subinterval (γ, β). Label the interval (gap) (α, β) as γ. Let us call it the root interval of the interval (gap) (α, β), label the right sub interval (right gap) of γ by (α, γ) at level $d + 1$ and the left sub interval (left gap) of γ by (γ, β) at level $d + 1$. The gaps are sequentially generated by $x_1, x_2, \cdots, x_n$, starting from $[0, x]$. We can construct the binary search tree from the above sequential bisection, connecting each pair of gaps, left gap (α, γ) and right gap (γ, β), with their root gap (α, β), and labeling the root gap as γ. As in the discrete model, let us call each gap "node". Let us also call each interval with length ≥ 1 "internal node" and each interval with length < 1 "external node". In Figure 6.1 we illustrate this incremental construction of the binary search tree. Then the binary search tree Figure 6.2 is obtained from the sequential bisection.

Our bisection process is described by a binary tree. The root $(0, x)$, is labeled as $X_{0,0}$. The node $(0, x_1)$ is labeled as $X_{1,0}$ and (x_1, x), as $X_{1,1}$. In

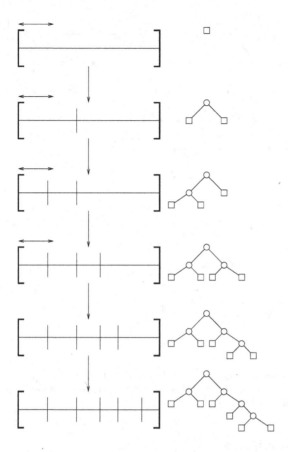

Fig. 6.1 The tree associated to a random sequential bisection. The final tree has $N_e(x,0) = 0$, $N_i(x,0) = 1$, $N_e(x,1) = 0$, $N_i(x,1) = 2$, $N_e(x,2) = 3$, $N_i(x,2) = 1$, $N_e(x,3) = 1$, $N_i(x,3) = 1$, $N_e(x,4) = 2$, $N_i(x,4) = 0$.

general, the node at the $(d-1)$-th level, $X_{d-1,j}$, if it is an internal node, has two child nodes which are labeled as $X_{d,2j}$ and $X_{d,2j+1}$, as in Figure 6.2. The external nodes correspond to remaining subintervals in the random sequential bisection, while the internal nodes correspond to intermediate subintervals to be eventually divided.

In the binary tree $T(x)$ associated with the random sequential bisection starting from $(0,x)$, let $N_i(x,d)$ and $N_e(x,d)$ denote the numbers of internal and external nodes at the d-th level respectively (see Figure 6.1). Let $m_i(x,d)$ and $m_e(x,d)$ denote their expected values respectively, which will be studied in Section 6.3.

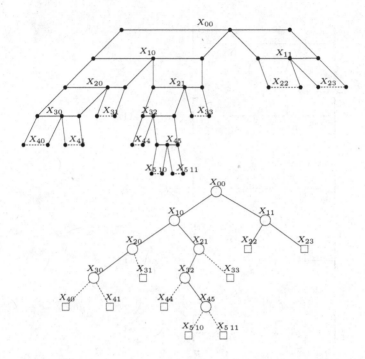

Fig. 6.2   Random sequential bisection and the associated binary tree. $X_{d,j}$: subinterval lengths/node labels. Circles are internal nodes ($X_{d,j} \geq 1$), squares are external nodes (children of internal nodes and $X_{d,j} < 1$)

6.2   Binary search tree

We note that the asymptotic shape of the binary search tree has been determined in [Robson (1979); Mahmoud and Pittel (1984); Pittel (1984)] and we will see that it is related with the random bisection model, though different from it. Our exposition closely follows [Mahmoud (1992)].

Suppose we want to store in a computer a set of data, whose elements can be compared by an ordering relation like $<$. The binary search tree for an input sequence of elements is constructed to store the date as follows. A root node is created for the first element of the data. Then subsequent elements are guided to the left or right subtree, according to whether they are less than the root label or not, where they are subjected recursively to the same treatment, until a unique insertion position is found.

As an example [Mahmoud (1992)], suppose we want to grow a binary search tree from the input sequence 4, 6, 5, 3, 1, 2, 7. Figure 6.3 illustrates the step-by-step realization of the tree. The first item 4 is placed in the root node (level 0). The next item of the input sequence is 6 and it goes to the right of the root because $6 > 4$ (level 1). When 5 comes along it is first compared with the label of the root, and as $5 > 4$ it goes to the right subtree and is compared with 6, and because $5 < 6$ it is guided to the left of 6 and that is where 5 is inserted (level 2). The process continues with the rest of the numbers until we end up with the final tree of Figure 6.3.

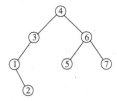

Fig. 6.3   Search tree for the sequence $4, 6, 5, 3, 1, 2, 7$

Binary search trees are constructed to satisfy the following search property: A binary search tree is a labeled binary tree that is either empty or has a labeled root node and (i) all the labels in the left subtree are less than the root label; (ii) all the labels in the right subtree are greater than the root label; (iii) the left and the right subtrees are also binary search trees.

A permutation $\Pi_n = \pi(1), \ldots, \pi(n)$ selected from the space of all $n!$ permutations of $\{1, \ldots, n\}$ will be called a *random permutation*, when all the permutations are equally likely. When the data stream is a sequence of n distinct keys from a totally ordered set and all shufflings of the n keys are equally likely, then obviously the sequence will be a random permutation. We assume that our data is a random permutation of $\{1, \ldots, n\}$. Note that *a priori* we do not know the size of the data.

Denote by $Y_{n,k}$ the random number of external nodes at level k, $k \geq 1$ of the tree. We look at U_n, the number of comparisons consumed by an unsuccessful search in a tree with n keys, to insert a new key of the data stream. Any new insertion will fall in one of the external nodes. Hence the probability $U_n = k$ is $Y_{n,k}/(n+1)$, the number of external nodes at level k.

Definition 6.1. The Stirling numbers of the first kind $\left[{n \atop k} \right]$ are defined as the number of permutations of n elements with k disjoint cycles. Another possible definition is to write the expansion

$$x(x-1)\ldots(x-n+1) = \sum_{k=0}^{n}(-1)^{n-k}\begin{bmatrix}n\\k\end{bmatrix}x^k$$

or

$$s(s+1)\cdots(s+n-1) = \sum_{k=0}^{\infty}\begin{bmatrix}n\\k\end{bmatrix}s^k.$$

They satisfy the following properties:

(1) $\begin{bmatrix}n\\n\end{bmatrix} = 1$ and $\begin{bmatrix}n\\1\end{bmatrix} = (n-1)!$.

(2) The recursion formula for $1 \le k \le n$:

$$\begin{bmatrix}n+1\\k\end{bmatrix} = n\begin{bmatrix}n\\k\end{bmatrix} + \begin{bmatrix}n\\k-1\end{bmatrix}.$$

(3) For each integer $k > 1$ the asymptotic expansions [Moser and Wyman (1958); Wilf (1995)]:

$$\frac{1}{(n-1)!}\begin{bmatrix}n\\k\end{bmatrix} = \gamma_1\frac{(\log n)^{k-1}}{(k-1)!} + \gamma_2\frac{(\log n)^{k-2}}{(k-2)!} + \cdots + \gamma_k + O\left(\frac{(\log n)^{k-2}}{n}\right) \tag{6.1}$$

where the γ_j are the coefficients in the expansion of the inverse of the Γ function:

$$\frac{1}{\Gamma(z)} = \sum_{j=1}^{\infty}\gamma_j z^j$$
$$= z + \gamma z^2 + \frac{5\gamma^2-\pi^2}{12}z^3 + \cdots,$$

where γ is the Euler constant.

Theorem 6.1. *[Lynch (1965)] We have for all $0 \le k \le n$:*

$$E[Y_{n,k}] = \frac{2^k}{n!}\begin{bmatrix}n\\k\end{bmatrix}.$$

Proof. We follow here the proof by [Mahmoud (1992)]. Conditioning on the event that the first integer in the permutation is i, or equivalently that the root label is i, the number of external nodes in the tree at level k is the same as the number of external nodes in the left subtree at level $k-1$ from the root of the left subtree, plus the number of external nodes in the right subtree at level $k-1$ from the root of the right subtree. Subject to the condition, the left subtree has $i-1$ nodes and the right subtree has $n-i$ nodes; therefore, the number of external nodes at level $k-1$ in the left subtree is $Y_{i-1,k-1}$ and at level $k-1$ in the right subtree is $Y_{n-i,k-1}$. Hence, we obtain the following equality

$$E[Y_{n,k} \mid i \text{ is in the root}] = E[Y_{i-1,k-1}] + E[Y_{n-i,k-1}].$$

Unconditioning we obtain

$$E[Y_{n,k}] = \sum_{i=1}^{n}(E[Y_{i-1,k-1}] + E[Y_{n-i,k-1}])Pr(i \text{ is in the root}).$$

The probability of the event that i is in the root is $1/n$. So,

$$nE[Y_{n,k}] = \sum_{i=1}^{n} E[Y_{i-1,k-1}] + \sum_{i=1}^{n} E[Y_{n-i,k-1}]$$

or

$$(n+1)E[Y_{n+1,k}] = \sum_{i=1}^{n+1} E[Y_{i-1,k-1}] + \sum_{i=1}^{n+1} E[Y_{n+1-i,k-1}].$$

Hence we have by subtraction

$$(n+1)E[Y_{n+1,k}] - nE[Y_{n,k}] = 2E[Y_{n,k-1}]. \tag{6.2}$$

If we write

$$E[Y_{n,k}] = \frac{2^k}{n!}S_{n,k}$$

then Equation (6.2) is transformed into

$$S_{n+1,k} = nS_{n,k} + S_{n,k-1},$$

which is the recurrence satisfied by Stirling's numbers of the first kind and has the same boundary conditions, because $Y_{0,0} = 1, Y_{0,k} = 0$, for $k > 0$, and $Y_{n,0} = 0$, for $n > 0$. Since

$$S_{n+1,k}s^k = nS_{n,k}s^k + S_{n,k-1}s^k,$$

for the generating function

$$g(n,s) = \sum_{k=0}^{\infty} S_{n,k}s^k$$

we have

$$g(n+1,s) = (n+s)g(n,s),$$

which means that

$$g(n,s) = s(s+1)\cdots(s+n-1).$$

It follows that

$$S_{n,k} = \begin{bmatrix} n \\ k \end{bmatrix},$$

by the definition of the Stirling number of the first kind.   □

If one takes a fixed k and consider the asymptotic of Stirling numbers from Equation (6.1), then we get

$$E[Y_{n,k}] \sim \frac{2^k}{n} \frac{(\log n)^{k-1}}{(k-1)!}.$$

We can make a discrete version of the random sequential bisection. We assume that our data is a random permutation of $\{1, \ldots, n\}$. Let a sequence $x_1, x_2, \cdots, x_n \in (0, x)$ of cars with length $l = 0$ be generated by the random permutation. Apply the above random sequential bisection starting from the interval $(0, n + 1)$.

Let us revisit the example ([Mahmoud (1992)] p. 58) to grow a binary search tree from the input sequence 4, 6, 5, 3, 1, 2, 7 and to understand the binary search tree by using the idea of bisection [Sibuya and Itoh (1987)]. The step-by-step realization of the tree is illustrated in Figure 6.3. We can state the above algorithms of binary search as a discrete sequential bisection which is a kind of discrete car parking of the case $d = 2$ given in Chapter 2 with $l = 0$ as introduced in Chapter 7. We introduce the discrete sequential bisection from the above input sequence 4, 6, 5, 3, 1, 2, 7. By the first item 4, the interval (gap) $(0, 8)$, at level 0, is bisected into the left gap (left subinterval), $(0, 4)$ and the right gap (right subinterval), $(4, 8)$, which are at level 1. Let us say that the gap $(0, 8)$ is the root of the left gap $(0, 4)$ and the right gap $(4, 8)$. The next item 6 of the sequence goes to the right gap $(4, 8)$ since $4 < 6$, and makes two gaps $(4, 6)$ and $(6, 8)$ which are at level 2. The gap $(4, 8)$ is the root gap of the left gap $(4, 6)$ and right gap $(6, 8)$. When the item 5 comes along it is in $(4, 8)$ at level 1, since $4 < 5$, and then guided to the gap $(4, 6)$ at level 2, since $5 < 6$, and makes gaps $(4, 5)$ and $(5, 6)$ at level 3. The gap $(4, 6)$ is the root gap of the left subgap $(4, 5)$ and right subgap $(5, 6)$. The process continues until all gaps have length lower than or equal to 1, which is shown by the tree of Figure 6.3.

We can reconstruct the original binary search tree from the above sequential bisection, connecting each pair of gaps, left gap (α, t) and right gap (t, β), with their root gap (α, β), and labeling the root gap as t. Then the binary search tree is obtained from the sequential bisection.

For the discrete sequential bisection, let us call each gap with length ≥ 1 "internal nodes" and the gap with length < 1 "external nodes", respectively.

6.3   Expected number of nodes at the d-th level

If the first division point of the interval $(0, x)$ is $Y, 0 < Y < x$, then $N_i(x, d) = N_i(Y, d-1) + N_i(x - Y, d-1)$, and the expectation of this equality shows that for $1 \leq x < \infty$ and $d = 1, 2, \cdots$.

$$\begin{aligned} m_i(x, d) &= \tfrac{1}{x} \int_0^x m_i(y, d-1) + m_i(x - y, d-1) dy \\ &= \tfrac{2}{x} \int_0^x m_i(y, d-1) dy \end{aligned} \tag{6.3}$$

with

$$m_i(x, 0) = \begin{cases} 0 & \text{if } 0 \leq x < 1, \\ 1 & \text{if } 1 \leq x < \infty. \end{cases} \tag{6.4}$$

By using Equations (6.3) and (6.4) and integrating we find:

$$m_i(x, 1) = \begin{cases} 0 & \text{if } 0 \leq x < 1, \\ \tfrac{2}{x}(x - 1) & \text{if } 1 \leq x < \infty, \end{cases}$$

$$m_i(x, 2) = \begin{cases} 0 & \text{if } 0 \leq x < 1, \\ \tfrac{2}{x}(x - 1 - \log x) & \text{if } 1 \leq x < \infty. \end{cases}$$

We can actually integrate the recursion relation (6.3) for $m_i(x, d)$ for all d and get:

$$m_i(x, d) = \begin{cases} 0 & \text{if } 0 \leq x < 1, \\ 2^d \tfrac{1}{x} \sum_{k=d}^{\infty} \frac{(\log x)^k}{k!} & \text{if } 1 \leq x < \infty, \end{cases} \tag{6.5}$$

for $d = 0, 1, 2, \cdots$.

In any binary tree, $N_i(x, d-1)$ internal nodes have $2N_i(x, d-1)$ child nodes, among which $N_i(x, d)$ are internal, therefore for $d = 1, 2, \cdots$

$$N_e(x, d) = 2N_i(x, d-1) - N_i(x, d). \tag{6.6}$$

The expectation of this equality shows that for $d = 1, 2, \cdots$,

$$m_e(x, d) = 2m_i(x, d-1) - m_i(x, d) = 2^d \frac{1}{x} \frac{(\log x)^{d-1}}{(d-1)!}. \tag{6.7}$$

The value of $m_e(x, 0)$ is undefined at present. Summarizing (6.5) and (6.7), we have proved the following results:

Theorem 6.2. *[Sibuya and Itoh (1987)] Among the possible 2^d nodes at the d-th level, $1 \leq d$, of the associated tree $T(x)$, the proportion of the expected number of the internal and the external nodes have the Poisson probabilities*

$$\frac{1}{x} \sum_{k=d}^{\infty} \frac{(\log x)^k}{k!} \quad \text{and} \quad \frac{1}{x} \frac{(\log x)^{d-1}}{(d-1)!}$$

respectively.

The implications of Theorem 6.2 are discussed in Section 6.5. But we can already state that $m_e(x,d)$ have the same asymptotic density as $E[Y_{n,k}]$ discussed before though they are not equal: $m_e(x,d)$ is irrational for d large while $E[Y_{n,k}]$ is always rational.

Note that it is possible to write a recursion relation for $m_e(x,d)$ similar to Equation (6.3):

$$m_e(x,d) = \begin{cases} 0 & \text{if } 0 \le x < 1, \\ \frac{2}{x} \int_0^x m_e(t, d-1)dt & \text{if } 1 \le x < \infty, \end{cases} \qquad (6.8)$$

for $d = 1, 2, \cdots$ which together with the initial condition

$$m_e(x,0) = \begin{cases} 1 & \text{if } 0 \le x < 1, \\ 0 & \text{if } 1 \le x < \infty \end{cases} \qquad (6.9)$$

allows one to find the desired expression of $m_e(x,d)$.

The numbers $m_e(x,d)$ and $m_i(x,d)$ satisfy the same integral equation but their initial conditions are different. For example the initial condition $m_e(x,d)$ vanishes in $(0,1)$ and the integration in (6.8) can be limited to the interval $(1,x)$, but not in (6.3). The initial condition (6.9) looks just like a conventional rule since an external node cannot appear at the roots of the subtrees.

6.4 Exponential distribution and uniform distribution

The distribution of a label $X_{d,j}$ of an external node, length of a remaining subinterval of random sequential bisection with the stopping rule, is the $(0,1)$ uniform distribution. To prove this, in the formal definition of $X_{d,j}$ in Section 6.1, let $j = (j_1 j_2 \cdots j_d)$ be the binary expression of a non negative integer index j, possibly having leading zeroes. Namely, $X_{d,j} = x \Pi_{k=1}^d U_{k(j_1 \cdots j_k)}$, where d is the smallest integer such that $x \Pi_{k=1}^d U_{k(j_1 \cdots j_k)}$ is less than one, or

$$-\sum_{k=1}^d \log U_{k(j_1 \cdots j_k)} > \log x.$$

The random variables $-\log U_{k(j_1 \cdots j_k)}$ are mutually independent standard exponential random variables. Since for the random variable with uniform distribution on $[0,1]$, we have the probability $Pr(U \le e^{-x}) = e^{-x}$, this gives

$$Pr(\log U \le -x) = Pr(-\log U \ge x) = e^{-x}. \qquad (6.10)$$

Let us show that the difference $-\sum \log U_{k(j_1 \cdots j_k)} - \log x = -\log X_{d,j}$ is a standard exponential variable under the above inequality condition and $X_{d,j}$ is conditionally a $(0,1)$ uniform random variable.

We will solve this problem by relating it to the waiting bus problem, whose analysis is done exhaustively in [Feller (1971)], Chapter 1, Sections 3 and 4. A classical model of bus arriving at random is by a Poisson random process, that is the probability that a bus arrives at a time t is

$$p(t) = \alpha e^{-\alpha t}.$$

Of course buses arrive one after the other and thus we have a function $g_n(t)$ which is the probability that n bus arrives before an instant t (see Section 4.6 and [Feller (1971)] for more details). Now the key question is what is the waiting time for a person waiting for a bus? It turns out that the probability of a waiting time τ is $p(\tau)$. The argument for this is sufficiently intricate that we prefer to refer the reader to the above reference that treats the problem extensively.

Clearly the probability $Pr(\log U \leq -x)$ is exactly a waiting time for a random Poisson process. In Section 7.5 we provide an analytic proof of this result, that follows [Sibuya and Itoh (1987)].

6.5 Asymptotic size of the associated tree

We study further total numbers $N_i(x) = \sum_{d=0}^{\infty} N_i(x, d)$ of the internal nodes and $N_e(x) = \sum_{d=1}^{\infty} N_e(x, d)$ of the external ones of the associated tree $T(x)$. By summing up (6.5) and (6.7) respectively, their expected values $m_i(x)$ and $m_e(x)$ are, if $1 \leq x < \infty$,

$$m_i(x) = \begin{cases} 2x - 1 \text{ if } x \geq 1, \\ 0 \text{ if } x < 1, \end{cases} \text{ and } m_e(x) = \begin{cases} 2x \text{ if } x \geq 1, \\ 1 \text{ if } x < 1. \end{cases}$$

We will give another proof of the formula for $m_i(x)$ in Theorem 7.1.

The relationship between these values comes also from the fact that $N_e(x) = N_i(x) + 1$ in any binary tree because of (6.6), and from the discussion at the end of the last section. Note that $m_i(x)$ is the solution to the integral equation

$$m_i(x) = \frac{2}{x} \int_0^x m_i(y) dy + 1, \text{ for } 1 \leq x < \infty \tag{6.11}$$

with

$$m_i(x) = 0, \text{ for } 0 \leq x < 1,$$

and $m_e(x)$ is the solution to

$$m_e(x) = \frac{2}{x} \int_0^x m_e(y)dy, \tag{6.12}$$

with the convention

$$m_e(x) = 1 \text{ if } 0 \le x < 1.$$

Equation (6.12) is obtained from (6.11) by putting $m_e(x) = m_i(x) + 1$.

Let $v_e(x)$ denote the variance of $N_e(x)$. For $2 \le x$, the variance of $N_i(x) = N_e(x) - 1$ is equal to

$$v_e(x) = (8 \log 2 - 5)x \simeq 0.54517744x, \text{ if } 2 \le x < \infty.$$

The proof is given in Chapter 7 following [Van Zwet (1978); Sibuya and Itoh (1987)]. Based on the linearity of $m_e(x)$ and $v_e(x)$ in x, we have the central limit theorem [Sibuya and Itoh (1987)] directly from the argument by [Dvoretzky and Robbins (1964)] in Chapter 5.

Theorem 6.3. *The standardized random variable*

$$\frac{N_e(x) - m_e(x)}{\sqrt{v_e(x)}}$$

is asymptotically normally distributed as $x \to \infty$.

6.6   Asymptotic shape of the associated tree

In Sections 6.3 and 6.5 some facts on the size of the associated tree $T(x)$ are shown. In this section its shape is discussed. Firstly we note that Theorem 6.2 actually shows the following fact: If the number of external nodes at a level of $T(x)$ is small, it means either that almost all nodes of the level are internal or that there are only a few internal nodes at the level.

Theorem 6.4. *[Sibuya and Itoh (1987)] As x and d increase to infinity satisfying $d = c \log x$,*

$$\left.\begin{array}{l} m_e(x, d) \\ m_i(x, d) \qquad \text{if } c > 1 \\ 2^d - m_i(x, y) \text{ if } c < 1 \end{array}\right\} = \frac{1}{\sqrt{2\pi d}} e^{-d\gamma(c)}(1 + Q(1/d)),$$

where $\gamma(c) = 1/c + \log(c/2) - 1$.

Proof. Let us first consider $m_e(x, d)$ and use its expression in Equation (6.7). Using Stirling's formula we obtain

$$m_e\left(\exp\left(\frac{d}{c}\right), d\right) = \frac{1}{\sqrt{2\pi d}}\exp\left\{-\gamma(c)d + O\left(\frac{1}{d}\right)\right\}.$$

Therefore $m_e(\exp(d/c), d)$ tends to 0 or ∞, if $\gamma(c) \geq 0$ or $\gamma(c) < 0$, respectively. The function $\gamma(c)$ decreases monotonically on $(0, 1]$ and increases monotonically on $[1, \infty[$ (see Figure 6.4).

Let us now consider the problem for $m_i(x, d)$. First of all let us take a Poisson probability $p(x; \mu)$:

$$p(x; \mu) = \frac{e^{-\mu}\mu^x}{x!}.$$

It is shown that its upper and lower tail probabilities are evaluated by the probabilities at the end of the tails [Barbour, Holst and Janson (1992)]; if $\mu - 1 < x$, since

$$\sum_{y=x}^{\infty}\frac{\mu^y}{y!} < \sum_{y=x}^{\infty}\frac{\mu^{y-x}}{(x+1)^{y-x}}\frac{\mu^x}{x!} = \frac{x+1}{x+1-\mu}\frac{\mu^x}{x!},$$

we have

$$p(x; \mu) < \sum_{y=x}^{\infty}p(y; \mu) < \frac{x+1}{x+1-\mu}p(x; \mu).$$

If $x < \mu$, we have

$$p(x; \mu) < \sum_{y=0}^{x}p(y : \mu) < \frac{x}{\mu - x}p(x; \mu).$$

Theorem 6.2 shows that the expected number of internal and external nodes at the d-th level satisfies

$$m_e(x, d+1) < 2^d - m_i(x, d) < \frac{d}{\log x - d}m_e(x, d), \text{ if } d < \log x,$$

and

$$m_e(x, d+1) < m_i(x, d) < \frac{d+1}{d+1-\log x}m_e(x, d+1), \text{ if } \log x < d+1.$$

Thus the asymptotic behavior of $m_e(x, d)$ determines that of $m_i(x, d)$. $\square$

Theorem 6.4 implies that

$$\lim m_e(x, d) = \lim m_i(x, d) = \begin{cases} 0 & \text{if } \bar{c} \leq c < \infty, \\ \infty & \text{if } 1 < c < \bar{c}, \end{cases}$$

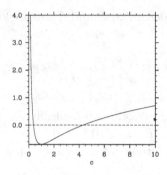

Fig. 6.4   The γ function

and

$$\lim m_e(x,d) = \lim 2^d - m_i(x,d) = \begin{cases} 0 & \text{if } 0 < c \leq \underline{c}, \\ \infty & \text{if } \underline{c} < c < 1, \end{cases}$$

where the limit means $d = c \log x \to \infty$, and $\bar{c} \simeq 4.31107041$ and $\underline{c} \simeq 0.373364616$ are solutions to $\gamma(c) = 0$. The function γ is plotted on Figure 6.4.

The limits $m_i(x,d) \to 0$ in the theorem mean that the probability of the following equivalent events approaches one:

$$A_u(x,d): \ N_i(x,d) = 0 \Leftrightarrow \max_j X_{d,j} < 1 \Leftrightarrow H(x) \geq d,$$

where $H(x)$ denotes the highest level of the external nodes of $T(x)$. Similarly, $2^d - m_i(x,d) \to 0$ means that with probability approaching one,

$$A_L(x,d): \ N_i(x,d) = 2^d \Leftrightarrow \min_j X_{d,j} > d \Leftrightarrow h(x) > d,$$

occurs, where $h(x)$ denotes the lowest level of the external nodes of $T(x)$. That is, the pair $H(x)$, $h(x)$ is $(4,2)$, $(5,2)$ and $(4,2)$ for Figures 6.1, 6.2 and 6.3, respectively.

Thus, the external nodes are located at levels between $\underline{c} \log x$ and $\bar{c} \log x$. In terms of $Z_{d,j} = X_{d,j}/x$, lengths of subintervals starting from $(0,1)$, this means the following:

Corollary 6.1. *With probability approaching one as $d \to \infty$,*

$$\frac{1}{\bar{c}} < \frac{-\log Z_{d,j}}{d} < \frac{1}{\underline{c}}.$$

6.7 More on the associated tree

In this associated binary tree, the probability $P_U(x, d)$ of the event $A_U(x, d)$ (see the discussion before Corollary of Theorem 6.4 in Section 6.6): $N_i(x, d) = 0 \Leftrightarrow \max_j X_{d,j} < 1 \Leftrightarrow H(x) \leq d$, satisfies the equation with P replaced by $P_U(x, d) = Pr(H(x) \leq d)$,

$$P(x, d) = \frac{1}{x} \int_0^x P(y, d-1)P(x-y, d-1)dy \qquad (6.13)$$

and the initial condition

$$P_U(x, 0) = \begin{cases} 1 & \text{if } 0 \leq x < 1, \\ 0 & \text{if } 1 \leq x < \infty. \end{cases}$$

The probability $P_L(x, d) = Pr(h(x) \geq d)$ of the dual event $A_L(x, d)$: $N_i(x, d) = 2^d \Leftrightarrow \min_j X_{d,j} \geq 1 \Leftrightarrow h(x) > d$, satisfies the same equation (6.13) with P replaced by P_L, and the initial condition

$$P_L(x, 0) = \begin{cases} 0 & \text{if } 0 \leq x < 1, \\ 1 & \text{if } 1 \leq x < \infty. \end{cases}$$

For the binary search tree, consider the highest level of the external node $\tilde{H}(n)$ and the lowest level of external node $\tilde{h}(n)$. The probabilities $\tilde{P}_U(n, d) = Pr(\tilde{H}(n) \leq d)$ and $\tilde{P}_L(n, d) = Pr(\tilde{h}(n) \geq d)$ satisfy a discrete analogue of (6.13).

$$\tilde{P}(n, d) = \frac{1}{n} \sum_{k=0}^{n-1} \tilde{P}(k, d-1)\tilde{P}(n-k-1, d-1)$$

with the initial conditions

$$\tilde{P}_U(n, 0) = \begin{cases} 1 & \text{if } n = 0, \\ 0 & \text{if } n = 1, 2, \cdots, \end{cases}$$

and

$$\tilde{P}_L(n, 0) = \begin{cases} 0 & \text{if } n = 0, \\ 1 & \text{if } n = 1, 2, \cdots \end{cases}$$

respectively. See [Robson (1979); Sibuya and Itoh (1987); Hattori and Ochiai (2006)] for further studies.

Chapter 7

The unified Kakutani Rényi model

We consider a sequential random packing process, which generalizes both Rényi's random packing problem, which we considered in Chapter 3 and Kakutani's interval splitting process which we considered in Chapter 6. Following [Komaki and Itoh (1992)] we consider the sequential random packing of cars of length $l \in [0, 1]$ into the interval $[0, x]$ where cars are put in intervals of length at least 1. If $l = 1$ we get Rényi's model while for $l = 0$ we get a rescaling of Kakutani's interval splitting procedure. Let us denote by $N_{x,l}$ the random variable of the number of cars put in the obtained sequential packing.

We first compute an integral expression of the limit random packing density $C_K(l)$ by using the Laplace transform. We then consider the case $l = 0$ of Kakutani's interval splitting procedure. We first prove the finiteness of the moments $E\left(N_{x,l}^m\right)$, a non-trivial result if $l = 0$. We compute explicitly the mean and variance of $N_{x,0}$ as a function of x.

Then we prove a central limit theorem for $N_{x,l}$ as $x \to \infty$ for any $l \in [0, 1]$. This is followed by the proof that $N_{x,l}/x$ converges almost surely to $C_K(l)$. A key instrument of the proof is a monotonicity theorem that is proved by using the tree structure introduced in Chapter 6.

In the final section we generalize the results of [Bankövi (1962)] to this unified Kakutani Rényi model. We give an integral expression for the number of gaps whose length is larger than $h \in [0, 1]$. This allows us to derive the probability density of the size of gaps and to reprove that it is uniform for Kakutani's interval splitting model. Then we compute the mean and variance of the size of gaps.

7.1   The limit random packing density

Cars of length, $l \in [0, 1]$ are allowed to park on the interval $[0, x]$ if there are spaces not less than 1 unit in length. Cars are sequentially parked, with the locations of the front of each car being uniformly distributed on $[0, x]$. The car actually parks at this site if it fits within a gap of length ≥ 1. If not, it resamples its preferred location. The procedure continues until none of the gaps have length greater than 1. Here, we study this model and obtain closed formulas for the limit packing densities and the limit distribution functions of gaps. We denote by $N_{x,l}$ the random variable of the number of cars put and we define

$$M_l(x) = E(N_{x,l})$$

the expected number of cars allowed to park on the interval $[0, x]$.

Lemma 7.1. *We have for $x \geq 1 - l$,*

$$M_l(x + l) = 1 + \tfrac{2}{x} \int_0^x M_l(y)dy \tag{7.1}$$

with $M_l(x) = 0$ for $0 \leq x < 1$.

Proof.   Let us condition on the initial position t of the first car that has been put in $[0, x + l]$. Of course we have $t \in [0, x]$. After this first car has been put we have two intervals of length t and $x - t$. Therefore we have the equality

$$E(N_{x+l,l}|t) = 1 + E(N_{t,l}) + E(N_{x-t,l}) = 1 + M_l(t) + M_l(x - t). \tag{7.2}$$

If one integrates the above equation then one gets

$$
\begin{aligned}
M_l(x + l) &= \tfrac{1}{x} \int_0^x E(N_{x+l,l}|t)dt \\
&= \tfrac{1}{x} \int_0^x 1 + M_l(t) + M_l(x - t)dt \\
&= 1 + \tfrac{2}{x} \int_0^x M_l(t)dt.
\end{aligned}
$$

The initial condition is easy to see. $\square$

By using this integral equation we get

$$
M_l(x) = \begin{cases}
0 & \text{if } 0 \leq x < 1, \\
1 & \text{if } 1 \leq x < 1 + l, \\
1 + \tfrac{2}{x}(x - l - 1) & \text{if } 1 + l \leq x < 1 + 2l.
\end{cases}
$$

Let us now introduce an expression that will show up many times in this chapter and which will allow to simplify expressions. Let us define for $t \in \mathbb{R}$

$$\Psi(t) = \exp\left(-2 \int_0^t \frac{1 - e^{-u}}{u} du\right).$$

Before stating the limit result, we give an integral equation that will show up several times in this chapter:

Lemma 7.2. *We have for $0 \leq l \leq 1$:*

$$\int_0^\infty \{1 + (1-l)t\}e^{-(1-l)t}\Psi(lt)dt = 2\int_0^\infty e^{-t}\Psi(lt)dt. \qquad (7.3)$$

Proof. First of all, one remarks that

$$\Psi(lt) = \exp\left\{-2\int_0^{lt}\frac{1-e^{-u}}{u}du\right\} = \exp\left\{-2\int_0^t\frac{1-e^{-lu}}{u}du\right\}.$$

If one sets

$$I(l) = \int_0^\infty \left(-e^{-t} + e^{-(1-l)t}\right)\Psi(lt)dt$$

then one gets after integration by part

$$\begin{aligned}
I(l) &= \int_0^\infty e^{-(1-l)t}(1 - e^{-lt})\Psi(lt)dt \\
&= -\int_0^\infty \{te^{-(1-l)t}\}'\frac{1}{-2}\Psi(lt)dt \\
&= \frac{1}{2}\int_0^\infty e^{-(1-l)t}(1 - (1-l)t)\Psi(lt)dt,
\end{aligned}$$

which after simplifications gives (7.3). □

If one sets $l = 1$ in Equation (7.3) then one gets the two expressions of C_R in formulas (3.13) and (3.17).

Theorem 7.1. *[Komaki and Itoh (1992)] The limit packing density of the cars of length $l \in [0,1]$ on the street of x in length is given by*

$$\lim_{x\to\infty} \frac{M_l(x)}{x} = C_K(l) \qquad (7.4)$$

with

$$C_K(l) = 2\int_0^\infty e^{-t}\Psi(lt)dt. \qquad (7.5)$$

Proof. Put $w_l(s) = e^s\int_0^\infty M_l(x)\exp(-sx)dx$. We have by the same method as for Theorem 3.1:

$$\frac{d}{ds}w_l(s) = \left\{(1-l) - \frac{2\exp(-ls)}{s}\right\}w_l(s) - \frac{1-l}{s} - \frac{1}{s^2}. \qquad (7.6)$$

The solution of Equation (7.6), using the boundary condition $\lim_{x\to\infty} w_l(s) = 0$, is

$$\begin{aligned}
w_l(s) = \frac{1}{s^2}\int_s^\infty &\{1 + (1-l)t\}\exp\{-(1-l)(t-s)\} \\
&\exp\left(-2\int_s^l\frac{1-\exp(-lu)}{u}du\right)dt.
\end{aligned}$$

By applying the Tauberian Theorem 3.2 and by a reasoning similar to the one of Theorem 3.3 for Rényi's packing problem, we get

$$\lim_{x \to \infty} \frac{M_l(x)}{x} = C_K(l)$$

with

$$C_K(l) = \int_0^\infty \{1 + (1-l)t\}e^{-(1-l)t}\Psi(lt)dt.$$

The conclusion follows from Lemma 7.2.   □

When $l = 1$, formula (7.4) coincides with the result obtained by [Rényi (1958)] in Theorem 3.3. When $l = 0$, we get $\lim_{x \to \infty} M_0(x)/x = 2$ obtained from (7.4) and it is actually possible to determine $M_0(x)$ in Theorem 7.1. The limit packing densities (7.5) are shown in Figure 7.1.

Note that Theorem 5.4 can be adapted to the generalized Kakutani's interval splitting and this yields:

$$V(N_{x+l,l}) \geq \frac{2}{x} \int_0^x V(N_{t,l})dt.$$

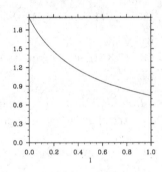

Fig. 7.1 Packing density for the unified Kakutani Rényi packing model in terms of $l \in [0,1]$

7.2 Expectation and variance of number of cars for $l = 0$

Remark that if $l > 0$ then by a simple packing argument we have

$$N_{x,l} \leq \frac{x}{l}$$

and thus the moments $E(N_{x,l}^m)$ are defined for all $m \in \mathbb{N}$. But $P(N_{x,0} > m) > 0$ for all $m \in \mathbb{N}$ and $x > 1$ and thus we have to prove that $E(N_{x,0}^m) < \infty$ in order to be able to continue the analysis.

Theorem 7.2. *[Van Zwet (1978)] For all $m > 0$ and $x \geq 0$ we have $E\left(N_{x,0}^m\right) < \infty$.*

Proof. Suppose first that $x < 2$. After the first car with $l = 0$ is parked, at most one of two spaces can have size greater than 1. So, when one leaves an interval, one never comes back to it. So, for all integer k we have

$$Pr(N_{x,0} > k) = Pr\left(x\Pi_{i=1}^k \max(U_i, 1 - U_i) > 1\right)$$

with mutually independent random variables $U_1, \ldots, U_k$ uniformly distributed on $[0, 1]$. Since the expectation of $\max(U_i, 1 - U_i)$ is $\frac{3}{4}$ we have

$$E\left(x\Pi_{i=1}^k \max(U_i, 1 - U_i)\right) = x\left(\frac{3}{4}\right)^k$$

and thus by Markov's inequality we get

$$Pr(N_{x,0} > k) \leq x\left(\frac{3}{4}\right)^k.$$

It follows that for any $m > 0$ we have

$$
\begin{aligned}
E(N_{x,0}^m) &= \sum_{k=0}^m k^m Pr(N_{x,0} = k) \\
&\leq \sum_{k=0}^m k^m Pr(N_{x,0} > k) \\
&\leq \sum_{k=0}^m k^m x\left(\frac{3}{4}\right)^k < \infty.
\end{aligned}
$$

Cars of length $l = 0$ are sequentially parked uniformly at random from the possible space until no interval is not less than 1. If $x < x'$ then each sequence, $I_1, I_2, \ldots, I_k$, of points generated by random sequential packing in $[0, x]$, can be scaled by $c = x'/x$ to a sequence of points $cI_1, cI_2, \ldots, cI_k$, in $[0, x']$, which can be further extended to a sequence $cI_1, cI_2, \ldots, cI_k, I_{k+1}, \ldots, I_{k'}$ with $k \leq k'$. Consequently $E(N_{x,0}^m) \leq E(N_{x',0}^m)$ for each $m > 0$, if the second integral exists.

At first step, for $0 < x < 2$ and $1 < y < 2$, let us consider the car parking with $l = 0$ in $[0, xy]$ and stops when all intervals have length at most x. We thus have $N_{y,0} + 1$ intervals of length at most x. Hence by the above argument $E(N_{x_j,0}^m) \leq E(N_{x,0}^m)$ for each $j = 1, 2, \ldots, k$.

Conditioning on getting the points $x_1, x_2, \ldots, x_k$ by the sequential car parking with $l = 0$, one gets for any $m > 0$ the inequality

$$E(N_{xy,0}^m) \leq \sum_{k=1}^{\infty} Pr(N_{y,0} + 1 = k)E\left((N_{x_1,0} + \cdots + N_{x_k,0})^m\right)$$

with mutually independent random variables $N_{x_i,0}$ with $i = 1, 2, \ldots, k$.

We have the equality

$$
\begin{aligned}
E\left((N_{x_1,0} + \cdots + N_{x_k,0})^m\right) &= \sum_{m_1+\cdots+m_k=m} \binom{m}{m_1,\ldots,m_k} E(N_{x_1,0}^{m_1} \cdots N_{x_k,0}^{m_k}) \\
&= \sum_{m_1+\cdots+m_k=m} \binom{m}{m_1,\ldots,m_k} \Pi_{i=1}^k E\left(N_{x_i,0}^{m_i}\right).
\end{aligned}
$$

By applying Hölder's inequality, we get

$$E(N_{x_1,0}^{m_1}) \leq \{E((N_{x_1,0}^{m_1})^{m/m_1})\}^{m_1/m} = E(N_{x_1,0}^m)^{m_1/m}.$$

Thus, since $E(N_{x_j,0}^m) \leq E(N_{x,0}^m)$ for $j = 1, 2, \ldots, k$, we get

$$E\left((N_{x_1,0} + \cdots + N_{x_k,0})^m\right) \leq \sum_{m_1+\cdots+m_k=m} \binom{m}{m_1,\ldots,m_k} E(N_{x,0}^m)$$
$$\leq k^m E(N_{x,0}^m)$$

and

$$E(N_{xy,0}^m) \leq \sum_{k=1}^{\infty} y \left(\tfrac{3}{4}\right)^{k-1} k^m E(N_{x,0}^m) < \infty.$$

At second step, take x for the above xy and take $1 < y < 2$ at the first step and apply the above argument. We recursively apply the argument and can prove that $E(N_{x,0}^m) < \infty$ for all $0 < m$ and for all $0 < x$. $\square$

Remind that $M_0(x) = E(N_{x,0})$ is the expected number of cars allowed to park on the interval $[0, x]$.

Proposition 7.1. *We have*

$$M_0(x) = \begin{cases} 0 \text{ if } x < 1, \\ 2x - 1 \text{ if } x \geq 1. \end{cases}$$

Proof. The result for $x < 1$ follows from the definition. We have for $1 \leq x$,

$$M_0(x) = 1 + \frac{2}{x} \int_0^x M_0(y) dy, \qquad (7.7)$$

with $M_0(x) = 0$ for $0 \leq x < 1$. Now $\sup_{y \geq t} M_0(y) < \infty$ for $t > 0$ because of (7.7) and hence (7.7) implies that M_0 is continuous and differentiable for $1 \leq x$. It also follows from (7.7) that

$$\lim_{x \to 1} M_0(x) = 1.$$

Thus $M_0(1) = 1$. Equation (7.7) can be differentiated and we get for $x \geq 1$ the differential equation

$$(xM_0(x))' = 1 + 2M_0(x).$$

The general solution of the equation is $M_0(x) = -1 + \alpha x$ and the value $M_0(1) = 1$ implies $\alpha = 2$. $\square$

We now consider the variance of $N_{x,0}$.

Theorem 7.3. *[Sibuya and Itoh (1987)] We have*

$$V(N_{x,0}) = \begin{cases} 0 \text{ if } 0 \leq x \leq 1, \\ 2 + 2x + 8x \log x - 4x^2 \text{ if } 1 \leq x \leq 2, \\ (8 \log 2 - 5)x \text{ if } x \geq 2. \end{cases}$$

Proof. For $0 \leq x < 1$, $V(N_{x,0}) = 0$, since $N_{x,0} = 0$. For $x = 1$, $V(N_{x,0}) = 0$ since $N_{x,0} = 1$ almost surely. For $1 \leq x \leq 2$, we obtain by using the expression of $M_0(t)$:

$$M_0(t) + M_0(x - t) + 1 - M_0(x) = \begin{cases} -2t + 1 & \text{if } 0 \leq t \leq x - 1, \\ 2 - 2x & \text{if } x - 1 \leq t \leq 1, \\ 2t - 2x + 1 & \text{if } 1 \leq t \leq x. \end{cases}$$

If one insert this into Equation (5.31) then we get

$$xV(N_{x,0}) = 2 \int_0^x V(N_{t,0})dt + \left(-\frac{2}{3} - 2x + 4x^2 - \frac{4x^3}{3} \right),$$

with $V(N_{1,0}) = 0$. Putting $U(x) = \int_0^x V(N_{t,0})dt$, we get an ordinary differential equation, which gives us

$$V(N_{x,0}) = 2 + 2x + 8x \log x - 4x^2. \tag{7.8}$$

Let us now consider the case $x \geq 2$. For that we use a trick that allows us to reduce the complexity of the computation. For $0 \leq x$ let us define

$$\nu(x) = 2x - 1 \text{ and } v(x) = E\left((N_{x,0} - \nu(x))^2 \right).$$

For $1 \leq x$, we have $M_0(x) = \nu(x)$ and $v(x) = V(N_{x,0})$. We have

$$\begin{aligned} v(x) &= \int_0^x \left\{ \sum_{n_1,n_2} (n_1 + n_2 + 1 - \nu(x))^2 Pr(N_{t,0} = n_1) \right. \\ &\quad \left. Pr(N_{x-t,0} = n_2) \right\} \times Pr(Y = t)dt \\ &= \int_0^x [\sum_{n_1,n_2} (n_1 + n_2 + 1 - \nu(t) - \nu(x - t) \\ &\quad + \nu(t) + \nu(x - t) - \nu(x))^2 \\ &\quad \times Pr(N_{t,0} = n_1)Pr(N_{x-t,0} = n_2)]Pr(Y = t)dt \\ &= \int_0^x [\sum_{n_1,n_2} \left((n_1 + n_2 - \nu(t) - \nu(x - t))^2 \right. \\ &\quad + 2(n_1 + n_2 - \nu(t) - \nu(x - t))(\nu(t) + \nu(x - t) + 1 - \nu(x)) \\ &\quad \left. + (\nu(t) + \nu(x - t) + 1 - \nu(x))^2 \right) \\ &\quad \times Pr(N_{t,0} = n_1)Pr(N_{x-t,0} = n_2)]Pr(Y = t)dt. \end{aligned}$$

We have the following equality:

$$\sum_{n_1,n_2} (n_1 + n_2 - \nu(t) - \nu(x - t))^2 Pr(N_{t,0} = n_1)Pr(N_{x-t,0} = n_2)$$

$$= \sum_{n_1} (n_1 - \nu(t))^2 Pr(N_{t,0} = n_1) + \sum_{n_2} (n_2 - \nu(x - t))^2 Pr(N_{x-t,0} = n_2)$$

$$+ 2 \sum_{n_1,n_2} (n_1 - \nu(t))(n_2 - \nu(x - t)) Pr(N_{t,0} = n_1)Pr(N_{x-t,0} = n_2)$$

$$= v(t) + v(x - t), \tag{7.9}$$

where

$$0 = \sum_{n_1,n_2} (n_1 - \nu(t))(n_2 - \nu(x-t)) Pr(N_{t,0} = n_1) Pr(N_{x-t,0} = n_2)$$
$$= \{\sum_{n_1} (n_1 - \nu(t)) Pr(N_{t,0} = n_1)\}$$
$$\{\sum_{n_2} (n_2 - \nu(x-t)) Pr(N_{x-t,0} = n_2)\},$$

since either $\sum_{n_1} (n_1 - \nu(t)) Pr(N_{t,0} = n_1)$ or $\sum_{n_2} (n_2 - \nu(x-t)) Pr(N_{x-t,0} = n_2)$ are 0 for $2 \leq x$. Also we have

$$\nu(t) + \nu(x-t) + 1 - \nu(x) = 0. \qquad (7.10)$$

Hence we have for $2 \leq x$

$$v(x) = V(N_{x,0}) = \frac{2}{x} \int_0^x v(t)dt, \qquad (7.11)$$

which gives for $2 \leq x < \infty$

$$V(N_{x,0}) = v(x) = V_{kat}x$$

with a constant V_{kat}. By considering the previous computation for $x = 2$ we get

$$V_{kat} = 8\log 2 - 5 \simeq 0.54517744$$

which is the required result.   □

The existence of the constant V_{kat} was proved in [Van Zwet (1978)] and its computation was done in [Sibuya and Itoh (1987)]. Another proof is available from [Pyke and Van Zwet (2004)]. We could have bypassed the above trick for $x \geq 2$ by directly integrating Equation (5.31) but this would have been more complicated. See Figure 7.2 for the Mean and Variance of $N_{x,0}$ as a function of x.

7.3   The central limit theorem

We can adapt the result of Chapter 5 and obtain estimate on the convergence rate and the central limit theorem for the generalized models.

Theorem 7.4. *Assume $0 < l \leq 1$:*

  (i) We have the convergence result

$$M_l(x) = C_K(l)x + lC_K(l) - 1 + O\left(\left(\frac{a}{x}\right)^{x-3/2}\right)$$

for some $a > 0$.

  (ii) We have the estimation of the variance

$$V(N_{x,l}) = \lambda_2(l)x + l\lambda_2(l) + O\left(\left(\frac{a}{x}\right)^{x-3/2}\right)$$

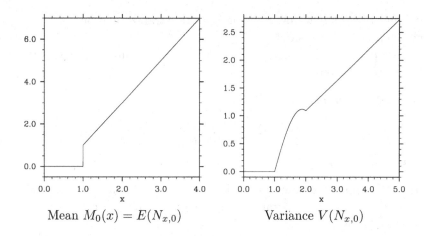

Mean $M_0(x) = E(N_{x,0})$ Variance $V(N_{x,0})$

Fig. 7.2   Mean and variance for $N_{x,0}$ in function of x

for some $\lambda_2(l)$ and $a > 0$.

(iii) The random variable

$$Z_{x,l} = \frac{N_{x,l} - M_l(x)}{\sqrt{V(N_{x,l})}}$$

is asymptotically normal $(0,1)$ as $x \to \infty$.

Proof.   If one does the change of variable $x = lx'$ and $M'(x') = M_l(lx')$ then Equation (7.1) becomes

$$M'(x' + 1) = 1 + \frac{2}{x'} \int_0^{x'} M'(y')dy' \qquad (7.12)$$

for $x' \geq \frac{1}{l} - l$ with the initial condition $M'(x') = 0$ for $x' \in [0, \frac{1}{l}]$. Equation (7.12) is the same as for Rényi's problem with the only exception that the initial conditions were different. But the initial conditions do not enter into the proof of convergence thus we get (i).

Let us now write

$$L_l(x) = C_K(l)x + lC_K(l) - 1 \text{ and } \phi_{k,l}(x) = E\left((N_{x,l} - L_l(x))^k\right).$$

The proof of (ii) then follows the one of Theorem 5.6. We first obtain

$$\phi_{2,l}(x + l) = \frac{2}{x} \int_0^x \phi_{2,l}(t)dt + \frac{2}{x} \int_0^x \phi_{1,l}(t)\phi_{1,l}(x - t)dt$$

and the estimation $\phi_{1,l} = O\left(\left(\frac{a}{x}\right)^{x-3/2}\right)$ for some $a > 0$. Then we apply the Corollary to Theorem 5.1 and we use $V(N_{x,l}) - \phi_{2,l}(x) = -(\phi_{1,l}(x))^2$ to get the result.

Statement (iii) follows by applying the same transformations and following the proof of Theorem 5.8.   □

Now we prove the central limit theorem for the Kakutani model again by the method of moments:

Theorem 7.5. *The random variable*

$$Z_{x,0} = \frac{N_{x,0} - M_0(x)}{\sqrt{V(N_{x,0})}}$$

is asymptotically normal $(0,1)$ *as* $x \to \infty$.

Proof. We apply the same methodology as in Theorem 5.8. Instead of having to use integral operator estimates, we simply solve differential equations. That is we define

$$\phi_k(x) = E\left((N_{x,0} - \nu(x))^k\right) \text{ with } \nu(x) = 2x - 1$$

and we need to prove the estimation

$$\phi_k(x) = c_k x^{\lfloor \frac{k}{2} \rfloor} + O\left(x^{\lfloor \frac{k}{2} \rfloor - 1 + \epsilon}\right). \tag{7.13}$$

Obviously we have $\phi_1(t) = 0$ for $t \geq 1$. The ϕ_k satisfy the recursion formula

$$\phi_m(x) = \frac{2}{x} \int_0^x \phi_m(t)dt + \frac{1}{x} \sum_{i=1}^{m-1} \binom{m}{i} \int_0^x \phi_i(t)\phi_{m-i}(x-t)dt.$$

Thus if one writes $F_m(x) = \int_0^x \phi_m(t)dt$ then one has

$$F'_m(x) = \frac{2}{x} F_m(x) + p_m(x)$$

with

$$p_m(x) = \frac{1}{x} \sum_{i=1}^{m-1} \binom{m}{i} \int_0^x \phi_i(t)\phi_{m-i}(x-t)dt.$$

The general solution of the differential equation is

$$\begin{cases} F_m(x) = \alpha_m x^2 + x^2 \int_{x_0}^x \frac{p_m(t)}{t^2} dt, \\ \phi_m(x) = 2\alpha_m x + p_m(x) + 2x \int_{x_0}^x \frac{p_m(t)}{t^2} dt, \end{cases} \tag{7.14}$$

where x_0 can be chosen arbitrarily. So, for $m = 3$ we get for $x \geq 3$:

$$\begin{aligned} p_3(x) &= \frac{2}{x} \int_0^x \phi_1(t)\phi_2(x-t)dt \\ &= \frac{2}{x} \int_0^1 (1 - 2t)\phi_2(x-t)dt \\ &= \frac{2}{x} \int_0^1 (1 - 2t)V_{kat}(x-t)dt \\ &= \frac{1}{x}V_{kat}\tfrac{1}{3}. \end{aligned}$$

Thus for $x \geq 3$ we have

$$F_3(x) = \alpha_3 x^2 - \frac{V_{kat}}{6}.$$

Thus $\phi_3(x) = 2\alpha_3 x$ for $x \geq 3$ as required. The proof of (7.13) for $m \geq 4$ of (7.13) follows closely the proof in Theorem 5.7 with the solution (7.14) replacing the integral equations. The coefficients c_k for k even are the right ones for applying the moment theorem and get the convergence to the Gaussian distribution. □

7.4 Almost sure convergence results

In order to prove the almost sure convergence result we will use the Borel-Cantelli Lemma and Markov inequality. The problem of such theorem is that we need a single probability space in order to apply them.

So, we will introduce a probability space which will contain the unified Kakutani Rényi model for all x and l. We have seen in Chapter 6 that the tree structure is an important tool. Thus we define an infinite binary tree $\mathcal{T}_\infty$ which starts from a vertex at level 0, which is connected to two vertices at level 1 and so on. More precisely let us denote by $X_{0,0}$ the top vertex. At level n we have 2^n vertices $X_{n,j}$ for $0 \leq j < 2^n$. For any n the vertex $X_{n,j}$ is connected to vertices $X_{n+1,2j}$ and $X_{n+1,2j+1}$.

The sample space that we define is the infinite product

$$\mathcal{P}_\infty = \Pi_{n=0}^\infty \left\{ \Pi_{j=0}^{2^n-1}[0,1] \right\}$$

with the probability being the product of the uniform density on each interval $[0,1]$. That is to every vertex $X_{n,j}$ of $\mathcal{I}_\infty$ we associate the interval $I_{n,j} = [0,1]$ and we define the probability on the infinite product of those intervals. A point in the sample space is denoted as $(t_{n,j})_{n \geq 0, 0 \leq j \leq 2^n-1}$ with $t_{n,j} \in [0,1]$.

Now, let us fix $x \geq 0$ and $l \in [0,1]$. We will define the unified Kakutani Rényi model at the level of the tree $\mathcal{I}_\infty$. Let us denote by $l_{n,j}(x,l)$ the length of the interval at the node $X_{n,j}$. It is a random variable on $\mathcal{P}_\infty$, thus it depends implicitly on the variables $t_{n,j}$ defining the sample space $\mathcal{P}_\infty$. We have $l_{0,0}(x,l) = x$. If $l_{0,0}(x,l) \geq 1$ then we can put a car of length l in $[0,x]$. The first position belongs to $[0,x-l]$ and the position is obtained from $t_{0,0} \in I_{0,0} = [0,1]$ by multiplying by $x-l$. Thus the interval is split into two intervals $[0, t_{0,0}(x-l)]$ and $[t_{0,0}(x-l)+l, x]$. Thus

$l_{1,0}(x,l) = t_{0,0}(x - l)$ and $l_{1,1}(x,l) = (1 - t_{0,0})(x - l)$. This can of course be generalized and we associate to every vertex $X_{n,j}$ a length $l_{n,j}$ with

$$\begin{cases} l_{n+1,2j}(x,l) = t_{n,j}(l_{n,j}(x,l) - l), \\ l_{n+1,2j+1}(x,l) = (1 - t_{n,j})(l_{n,j}(x,l) - l). \end{cases} \tag{7.15}$$

The above equation make sense if $l_{n,j}(x,l) \geq 1$. Otherwise we set $l_{n+1,2j}(x,l) = l_{n+1,2j+1}(x,l) = 0$.

For a given x, l and $p \in \mathcal{P}$ we have a non-extensible sequential random packing according to unified Kakutani Rényi model. In that context the number $N_{x,l}$ is simply

$$N_{x,l} = \sum_{n=0}^{\infty} |j \in \{0, \ldots, 2^n - 1\} \text{ s.t. } l_{n,j}(x,l) \geq 1|. \tag{7.16}$$

Most importantly the probability distribution of $N_{x,l}$ is actually the same as the one used in the definition of the unified Kakutani Rényi model. There is a small difference in the packing defined by the tree structure and a packing defined by the sequential random packing. In the sequential random packing, we know exactly when the cars are put and we have a sequence. In the tree structure, once we have put a car, the interval is split and we do not know if we put the next car in the first subtree or the second one. This is actually not important for the probability density. The integral relation between $N_{x+l,l}$, $N_{t,l}$, $N_{x-t,l}$ of Equation (7.2) corresponds to the fact that the infinite tree T_∞ is decomposed into two infinite trees and a vertex.

For a given x, l and a non-extensible random packing, we can reconstruct some of the positions $t_{n,j} \in [0,1]$ describing the sample space $\mathcal{P}_\infty$ but not all. Thus we can change the value of x and still define a packing. The problem is that this packing might be extensible and we cannot describe the extension if we do not know some variables $t_{n,j}$. See in Figure 7.3 the extension from $[0,3]$ to $[0,4]$ for $l = 1$. From this figure, it appears that the result of the extension operation depends on the tree structure.

Let us consider Equation (7.16) for a given set of $t_{n,j}$ in the above model.

Theorem 7.6. *We have the monotonicity properties*

$$N_{x,l} \leq N_{x',l} \text{ if } x \leq x'$$

and

$$N_{x,l} \geq N_{x,l'} \text{ if } l \leq l'.$$

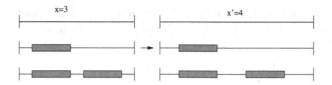

Fig. 7.3   The extension of a sequential random packing of $[0,3]$ to $[0,4]$ for $l = 1$

Proof.   If x increases then $l_{0,0}(x,l)$ increases and Equation (7.15) shows clearly that $l_{n,j}(x,l)$ increases for all n and j. Thus $N_{x,l}$, which is given by expression (7.16) increases as well. The proof of monotonicity in l is done in the same way.   □

This theorem implies in particular that $E\left(N_{x,l}\right)$ is non-decreasing. That is Proposition 3.1 is a direct consequence of the above theorem. It also implies the monotonicity of $C_K(l)$, which is apparent in Figure 7.1 and formula (7.5). It is interesting to consider the expansion factor $c_{n,j}$ when one goes from x to $x' > x$ in the sequential random packing. The position of the first car in the interval is multiplied by the factor $c_{0,0} = \frac{x'-l}{x-l}$. In the packing in the subinterval we can find the multiplicative factor. We have

$$c_{n+1,2j} = \frac{l_{n,j}(x',l) - l}{l_{n,j}(x,l) - l} = \frac{\cdot l_{n,j}(x,l)c_{n,j} - l}{l_{n,j}(x,l) - l}.$$

Thus the scaling factor is always greater than 1 but decreases in the tree except if $l = 0$, i.e. Kakutani's case in which case it is always x'/x.

Theorem 7.7. *For any $0 \le l \le 1$ almost surely*

$$\lim_{x \to +\infty} \frac{N_{x,l}}{x} = C_K(l).$$

Proof.   We know that $V(N_{x,l}) = O(x)$. Thus we have by Tchebychev inequality for $x = m^2$ the relation:

$$Pr\left(\left|N_{m^2,l} - E(N_{m^2,l})\right| > m^{5/3}\right) \le \frac{V(N_{m^2,l})}{m^{10/3}} = O\left(\frac{1}{m^{4/3}}\right).$$

Since the series $\frac{1}{m^{4/3}}$ converges we can apply the Borel-Cantelli lemma and conclude that almost surely $|N_{m,l} - E(N_{m,l})| \le m^{5/3}$ except, possibly, for a finite number of $m \in \mathbb{N}$. Thus we get the following almost sure result:

$$\limsup_{m} m^{-5/3}\left|N_{m^2,l} - E(N_{m^2,l})\right| \le 1.$$

This gives then the almost sure convergence

$$\lim_{m \to \infty} \frac{N_{m^2,l}}{E(N_{m^2,l})} = 1. \tag{7.17}$$

Now for any x let us denote by $m(x)$ the largest integer such that $m(x)^2 \leq x$. We thus have $m(x)^2 \leq x < (m(x) + 1)^2$ and for a given set of $t_{n,j}$ by Theorem 7.6 we have:

$$N_{m(x)^2,l} \leq N_{x,l} \leq N_{(m(x)+1)^2,l}.$$

By division this implies

$$\frac{N_{x,l}}{E(N_{x,l})} \leq \frac{N_{(m(x)+1)^2,l}}{E\left(N_{(m(x)+1)^2,l}\right)} \frac{E\left(N_{(m(x)+1)^2,l}\right)}{E(N_{x,l})}$$

$$\leq \frac{N_{(m(x)+1)^2,l}}{E\left(N_{(m(x)+1)^2,l}\right)} \frac{E\left(N_{(m(x)+1)^2,l}\right)}{E\left(N_{m(x)^2,l}\right)}$$

and by doing the same operation on the left we get

$$\frac{N_{m(x)^2,l}}{E\left(N_{m(x)^2,l}\right)} \frac{E\left(N_{m(x)^2,l}\right)}{E\left(N_{(m(x)+1)^2,l}\right)} \leq \frac{N_{x,l}}{E\left(N_{x,l}\right)}$$

$$\leq \frac{N_{(m(x)+1)^2,l}}{E\left(N_{(m(x)+1)^2,l}\right)} \frac{E\left(N_{(m(x)+1)^2,l}\right)}{E\left(N_{m(x)^2,l}\right)}.$$

The conclusion follows from the asymptotic estimation $E(N_{x,l}) = C_K(l)x + o(x)$ when $x \to \infty$, which implies

$$\lim_{x \to \infty} \frac{E\left(N_{(m(x)+1)^2,l}\right)}{E\left(N_{m(x)^2,l}\right)} = 1$$

and the almost sure convergence (7.17).   □

7.5   The limit distribution of a randomly chosen gap

The number of interval put in the unified Kakutani Rényi model is the random variable $N_{x,l}$. Therefore since the length of the interval is l, we are actually packing in $[0, x]$ the intervals $[t_1, t_1 + l]$, $[t_2, t_2 + l]$, ..., $[t_{N_{x,l}}, t_{N_{x,l}} + l]$ where the intervals $[t_i, t_i + l]$ are put sequentially. We denote by $\{t'_1, t'_2, \ldots, t'_{N_{x,l}}\}$ the reordering of the initial positions $\{t_1, t_2, \ldots, t_{N_{x,l}}\}$.

The numbers

$$I_x^{(1)} = t'_1, \quad I_x^{(k)} = t'_k - (t'_{k-1} + l) \text{ for } k = 2, 3, \ldots, N_{x,l}, \quad I_x^{(N_{x,l}+1)} = x - (t'_{N_{x,l}} + l)$$

will be called "gaps" of the sequential random packing of the unified Kakutani Rényi model. We want to study the average length of a gap.

Let us denote by $\vartheta_{x,l}(h)$ the random variable of the number of gaps larger than h. Then let us define

$$\begin{cases} \xi_{x,l}(h) = N_{x,l} + \vartheta_{x,l}(h), \\ M_{l,h}(x) = E(N_{x,l}) + E(\vartheta_{x,l}(h)) \end{cases}$$

so that $M_{l,h}(x)$ is the sum of the expected number of intervals and the expected number of gaps which are larger than h.

Lemma 7.3. *The function $M_{l,h}(x)$ satisfies the functional equation.*

$$M_{l,h}(x + l) = 1 + \frac{2}{x} \int_0^x M_{l,h}(t) \, dt \text{ for } x \geq 1 - l \qquad (7.18)$$

and the initial condition

$$M_{l,h}(x) = \begin{cases} 0 & \text{if } 0 \leq x < h, \\ 1 & \text{if } h \leq x \leq 1. \end{cases} \qquad (7.19)$$

Proof. We know that for the number $M_l(x) = E(N_{x,l})$ we have

$$M_l(x + l) = 1 + \frac{2}{x} \int_0^x M_l(t) dt.$$

Now let us condition on the position t of the first car that has been put into $[0, x + l]$. The interval $[0, x + l]$ is then subdivided into the interval $[0, t]$ and $[t + l, x + l]$ thus we have the equality

$$E(\vartheta_{x+1,l}(h)|t) = E(\vartheta_{t,l}(h)) + E(\vartheta_{x-t,l}(h))$$

and the equation follows by simple integration and remarking that the position t is uniformly chosen.

The initial condition (7.19) obviously follows from the model. $\qquad\square$

By integrating Equation (7.18) and using the initial condition (7.19) we get for $x \in [1, 1 + l]$

$$M_{l,h}(x) = \begin{cases} 1 & \text{if } h + l \geq x, \\ 3 - \frac{2h}{x-l}, & \text{if } h + l < x. \end{cases}$$

Let us now define

$$w_{l,h}(s) = \exp(ls) \int_1^\infty M_{l,h}(x) \exp(-sx) dx.$$

We have the following results:

Theorem 7.8. *For $0 \leq l \leq 1$ and $0 \leq h \leq 1$ we have:*

(i) The function $w_{l,h}(s)$ satisfies the equation

$$\frac{d}{ds} w_{l,h} = -\frac{2\exp(-sl)}{s} w_{l,h} - \frac{2}{s} \int_{1-l}^1 M_{l,h}(x) \exp(-sx) dx \\ - \frac{1-l}{s} \exp((l-1)s) M_{l,h}(1) - \frac{\exp((l-1)s)}{s^2}. \qquad (7.20)$$

(ii) The function $w_{l,h}$ has the expression

$$w_{l,h}(t) = \frac{1}{s^2} e^{2\int_0^s \frac{1-e^{-lu}}{u} du} \int_t^\infty \left(2s \int_{1-l}^1 M_{l,h}(x) \exp(-sx) dx + \right. \\ \left. \exp((l-1)s) \{s(1-l)M_{l,h}(1) + 1\} \right) \Psi(ls) ds. \qquad (7.21)$$

(iii) We have the limit

$$\lim_{x \to \infty} \frac{M_{l,h}(x)}{x} = C_K(l,h)$$

with

$$C_K(l,h) = \int_0^\infty \left[2t \int_{1-l}^1 M_{l,h}(x) \exp(-tx) dx \right] \Psi(lt) dt \\ + \int_0^\infty \{(1-l)M_{l,h}(1)t + 1\} \exp(-(1-l)t) \Psi(lt) dt. \tag{7.22}$$

Proof. The function $w_{l,h}$ can be rewritten as

$$w_{l,h}(s) = \int_{1-l}^\infty M_{l,h}(x+l) \exp(-sx) dx.$$

The function $M_{l,h}(x)$ is differentiable at $x = x_0$ if $x_0 > 1 - l$. Therefore if one multiplies by x Equation (7.18) and differentiate then one gets

$$x M_{l,h}'(x+l) + M_{l,h}(x+l) = 1 + 2M_{l,h}(x).$$

Multiplying by e^{-sx} and integrating we get

$$\int_{1-l}^\infty x M'(x+l) e^{-sx} dx + w_{l,h}(s) = \frac{1}{s} e^{-s(1-l)} + 2 \int_{1-l}^\infty M_{l,h}(x) e^{-sx} dx.$$

Then by integration by part we get

$$\int_{1-l}^\infty x M_{l,h}'(x+l) e^{-sx} dx = -\frac{d}{ds} \left(\int_{1-l}^\infty M_{l,h}'(x+l) e^{-sx} dx \right)$$
$$= -\frac{d}{ds} \left([M_{l,h}(x+l) e^{-sx}]_{1-l}^\infty \right.$$
$$\left. - \int_{1-l}^\infty M_{l,h}(x+l)(-s) e^{-sx} dx \right)$$
$$= -\frac{d}{ds} \left(-M_{l,h}(1) e^{-s(1-l)} + s w_{l,h}(s) \right).$$

By combining above equations we get

$$\frac{1}{s} e^{-s(1-l)} + 2 \int_{1-l}^1 M_{l,h}(x) e^{-sx} dx + 2 e^{-ls} w_{l,h}(s) \\ = -M_{l,h}(1)(1-l) e^{-s(1-l)} - s \frac{dw_{l,h}}{ds}(s) - w_{l,h}(s) + w_{l,h}(s).$$

By simplifying we get Equation (7.20).

This equation can be rewritten as

$$\frac{dw_{l,h}}{ds}(s) = w_{l,h}(s) \left[-2 \frac{\exp(-ls)}{s} \right] + H(s)$$

with

$$H(s) = -\frac{2}{s} \int_{1-l}^1 M_{l,h}(x) \exp(-sx) dx \\ - \frac{1-l}{s} \exp((l-1)s) M_{l,h}(1) - \frac{\exp((l-1)s)}{s^2}.$$

The solution of the homogeneous equation is

$$w_0(s) = \frac{1}{s^2} e^{2 \int_0^s \frac{1-e^{-lu}}{u} du}.$$

If one writes $w_{l,h}(s) = w_0(s)h(s)$ then one gets

$$-\frac{dh}{ds}(s) = \left(2s \int_{1-l}^{1} M_{l,h}(x) \exp(-sx)dx + \exp((l-1)s)\{s(1-l)M_{l,h}(1) + 1\}\right)\Psi(ls).$$

This derivative is integrable so one may write

$$h(s) = \alpha - \int_s^\infty \frac{dh}{ds}(u)du$$

and so

$$w_{l,h}(s) = \alpha w_0(s) - w_0(s)\int_s^\infty \frac{dh}{ds}(u)du.$$

We have the bound

$$\begin{aligned}
0 \le M_{l,h}(x) &= E(N_{x,l}) + E(\vartheta_{x,l}(h)) \\
&\le E(N_{x,l}) + 1 + E(N_{x,l}) \\
&\le 1 + 2E(N_{x,l}) \\
&\le 1 + 2(2x + o(x)),
\end{aligned}$$

where we have used Theorem 7.1 and the bound $C_K(l) \le 2$ for the packing density, which follows easily from Theorem 7.1.

Therefore $M_{l,h}(x) = 1 + O(x)$ from which one gets the estimation

$$w_{l,h}(s) = O\left(e^{(l-1)s}\frac{1}{s}\right) \text{ for } s \gg 1.$$

Since $l \le 1$ we have thus $w_{l,h}(s) = o(1)$ and since $\lim_{s\to\infty} w_0(s) = 1$ this implies $\alpha = 0$ and formula (7.21).

Statement (iii) follows in the same way as Theorem 3.3 by application of the Tauberian theorem 3.2. $\qquad\square$

The above result, when specialized to the case $l = 1$ gives the result by [Bankövi (1962)]:

Corollary 7.1. *[Bankövi (1962)] (i) For $0 < h \le 1$, we have the limit*

$$\lim_{x\to+\infty} \frac{M_{1,h}(x)}{x} = C_B(h)$$

with

$$C_B(h) = 2\int_0^\infty \exp\left\{-ht - 2\int_0^t \frac{1-e^{-u}}{u}du\right\}dt. \qquad (7.23)$$

(ii) For the average number $E(\vartheta_{x,1}(h))$ of gaps we have

$$\lim_{x\to\infty} \frac{E(\vartheta_{x,1}(h))}{x} = 2\int_0^\infty \left\{e^{-ht} - e^{-t}\right\}\exp\left\{-2\int_0^t \frac{1-e^{-u}}{u}du\right\}dt. \qquad (7.24)$$

Proof. When substituting $l = 1$ in formula (7.22) we get:

$$\lim_{x \to +\infty} \frac{M_{1,h}(x)}{x} = C(1,h)$$
$$= \int_0^\infty \left[1 + 2t \int_0^1 M_{1,h}(x) \exp(-tx) dx \right] \Psi(t) dt$$
$$= \int_0^\infty \left[2e^{-th} - 2e^{-t} + 1 \right] \Psi(t) dt$$
$$= 2 \int_0^\infty e^{-th} \Psi(t) dt$$

where we have used both expressions of C_R from formula (3.16).

So, (i) holds and (ii) follows by simple subtraction. $\qquad\square$

We can now consider the problem of estimating the average length of gaps. Let us denote by $I_{x,l}$ the length of a randomly chosen gap. We have the following theorem:

Theorem 7.9. *[Komaki and Itoh (1992)] The limiting distribution function of $I_{x,l}$ as $x \to \infty$ is given by*

$$\lim_{x \to \infty} Pr(I_{x,l} \leq h) = G_l(h) \qquad (7.25)$$

with

$$G_l(h) = \begin{cases} 2 - \dfrac{F_1(h,l)}{C_K(l)} & \text{if } h + l \geq 1 \\ h \dfrac{F_2(l)}{C_K(l)} & \text{if } h + l < 1 \end{cases}$$

and

$$\begin{cases} F_1(h,l) = 2 \int_0^\infty \exp(-ht) \Psi(lt) dt \\ F_2(l) = 2 \int_0^\infty t \exp(-(1-l)t) \Psi(lt) dt. \end{cases}$$

Proof. The quotient

$$\frac{\vartheta_{x,l}(h)}{N_{x,l} + 1}$$

is the ratio of the number of gaps not smaller than h, relative to the number of all gaps. The equality

$$Pr(I_{x,l} \geq h) = E\left(\frac{\vartheta_{x,l}(h)}{N_{x,l} + 1} \right)$$

obviously follows from the given definitions. Let us write the random variable

$$\rho_{x,l}(h) = \frac{\xi_{x,l}(h) + 1}{N_{x,l} + 1}.$$

We have

$$\frac{\vartheta_{x,l}(h)}{N_{x,l} + 1} = \rho_{x,l}(h) - 1.$$

Then we get successively

$$Pr(I_{x,l} \le h) = 1 - Pr(I_{x,l} > h)$$
$$= 1 - E\left(\frac{\vartheta_{x,l}(h)}{N_{x,l}+1}\right)$$
$$= 1 - (E(\rho_{x,l}(h)) - 1) = 2 - E(\rho_{x,l}(h)).$$

From Theorem 7.8 we have estimation on $E(\xi_{x,l}(h))$ and $E(N_{x,l})$ but we cannot use directly those estimations for the expectation of the quotient. Thus something else is needed, that is we will prove that the random variables $\xi_{x,l}(h)$ and $N_{x,l}$ are well approximated by their means.

We know from Theorem 7.4 (ii) that $V(N_{x,l}) = O(x)$ and by using the same technique one can prove easily that $V(\xi_{x,l}(h)) = O(x)$.

Let us write $\sigma_l(x) = \sigma(N_{x,l})$ and $\sigma_{l,h}(x) = \sigma(\vartheta_{x,l}(h))$. Let A and $A(h)$ denote the events

$$|N_{x,l} - M_l(x)| \le \lambda_1 \sigma_l(x)$$

and

$$|\xi_{x,l}(h) - M_{l,h}(x)| \le \lambda_2 \sigma_{l,h}(x),$$

respectively, where λ_1 and λ_2, are fixed and arbitrarily chosen positive numbers; $\overline{A}$ and $\overline{A(h)}$ denote the complements of A and $A(h)$, respectively. According to Tchebychev inequality

$$Pr\left(\overline{A}\right) < \frac{1}{\lambda_1^2} \text{ and } Pr\left(\overline{A(h)}\right) < \frac{1}{\lambda_2^2}$$

hold. Since $M_l(x)$, $M_{l,h}(x) = O(x)$ and $\sigma_l(x)$, $\sigma_{l,h}(x) = O(\sqrt{x})$ there exists a number x_0 such that for $x > x_0$ we have

$$M_l(x) + 1 > \lambda_1 \sigma_l(x) \text{ and } M_{l,h}(x) + 1 > \lambda_2 \sigma_{l,h}(x).$$

The variable $\rho_{x,l}(h)$ can be written in the form

$$\rho_{x,l}(h) = \frac{\xi_{x,l}(h) - M_{l,h}(x) + M_{l,h}(x) + 1}{N_{x,l} - M_l(x) + M_l(x) + 1}$$

thus, for $x > x_0$ we obtain the estimate

$$1 - \frac{1}{\lambda_1^2} - \frac{1}{\lambda_2^2} < Pr\left(A \cap A(h)\right) < Pr(a < \rho_{x,l}(h) < b) \tag{7.26}$$

with

$$a = \frac{M_{l,h}(x)\left(1 + \frac{1-\lambda_2 \sigma_{l,h}(x)}{M_{l,h}(x)}\right)}{M_l(x)\left(1 + \frac{1+\lambda_1 \sigma_l(x)}{M_l(x)}\right)} \text{ and } b = \frac{M_{l,h}(x)\left(1 + \frac{1+\lambda_2 \sigma_{l,h}(x)}{M_{l,h}(x)}\right)}{M_l(x)\left(1 + \frac{1-\lambda_1 \sigma_l(x)}{M_l(x)}\right)}.$$

Considering the asymptotic behavior of the functions $M_l(x)$, $M_{l,h}(x)$, $\sigma_l(x)$ and $\sigma_{l,h}(x)$ (7.26) can be written in the form

$$Pr\left(\frac{M_{l,h}(x)}{M_l(x)}(1-\epsilon_1) < \rho_{x,l}(h) < \frac{M_{l,h}(x)}{M_l(x)}(1+\epsilon_2)\right) > 1 - \frac{1}{\lambda_1^2} - \frac{1}{\lambda_2^2}, \quad (7.27)$$

where $\epsilon_2 = \epsilon_2(x; h, \lambda_1, \lambda_2)$, $\epsilon_1(x; h, \lambda_1, \lambda_2)$, $|\epsilon_1| \to 0$ and $|\epsilon_2| \to 0$ as $x \to +\infty$. It follows from (7.27) that

$$Pr\left(\left|\rho_{x,l}(h) - \frac{M_{l,h}(x)}{M_l(x)}\right| > \frac{M_{l,h}(x)}{M_l(x)}\max(|\epsilon_1|,|\epsilon_2|)\right) < \frac{1}{\lambda_1^2} + \frac{1}{\lambda_2^2}.$$

Considering the fact that $1 \le \rho_{x,l}(h) \le 2$ this means that

$$\lim_{x\to\infty} E\left(\rho_{x,l}(h)\right) = \frac{\lim_{x\to\infty}\frac{M_{l,h}(x)}{x}}{\lim_{x\to\infty}\frac{M_l(x)}{x}} = \frac{C_K(l,h)}{C_K(l)}.$$

One can remark that $C_K(l) = C_K(l,1)$ since $\xi_{x,l}(1) = N_{x,l}$ up to a set of measure 0.

We now need to evaluate the integral expressing $C_K(l, h)$. If $h + l \ge 1$ then we have

$$\begin{aligned}
C_K(l,h) &= \int_0^\infty \left[2t\int_h^1 \exp(-tx)dx\right]\Psi(lt)dt \\
&+ \int_0^\infty \{(1-l)t + 1\}\exp(-(1-l)t)\Psi(lt)dt \\
&= \int_0^\infty \left[2e^{-th} - 2e^{-t} + (1-l)te^{-(1-l)t} + e^{-(1-l)t}\right]\Psi(lt)dt.
\end{aligned}$$

By using Lemma 7.2 this simplifies to

$$C_K(l,h) = 2\int_0^\infty e^{-th}\Psi(lt)dt$$

and then gives the required result. If $h + l < 1$ then we get

$$\begin{aligned}
C_K(l,h) &= \int_0^\infty \left[2t\int_{1-l}^1 \exp(-tx)dx + ((1-l)M_{l,h}(1)t + 1)\exp(-(1-l)t)\right] \\
&\quad \Psi(lt)dt \\
&= \int_0^\infty \left[2e^{-(1-l)t} - 2e^{-t} + 3(1-l)te^{-(1-l)t} + e^{-(1-l)t}\right]\Psi(lt)dt \\
&\quad - 2h\int_0^\infty te^{-(1-l)t}\Psi(lt)dt \\
&= \int_0^\infty 4e^{-t}\Psi(lt)dt - 2h\int_0^\infty te^{-(1-l)t}\Psi(lt)dt \\
&= 2C_K(l) - 2h\int_0^\infty te^{-(1-l)t}\Psi(lt)dt.
\end{aligned}$$

This implies the required result.   □

When one specializes to the case $l = 1$ then we get the result by [Bankövi (1962)]

$$\lim_{x\to\infty} Pr(I_{x,1} < h) = G(h) = 2 - C_R^{-1}C_B(h)$$

with $C_B(h)$ from Equation (7.23). Evaluated value of $G(h)$ are shown in Figure 7.4.

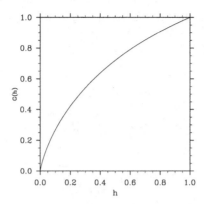

Fig. 7.4 The limit probability $G(h)$ as a function of h

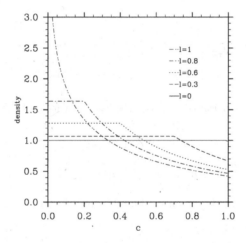

Fig. 7.5 The probability densities of the unified Kakutani Rényi model as a function of l

The limit density of $I_{x,l}$ for several values of l are illustrated in Figure 7.5. When $l = 0$, (7.25) gives $\lim_{x \to \infty} Pr(I_{x,l} \le h) = h$ that is the density is uniform. We have already seen this result in Section 6.4.

Let us now evaluate the moments of $I_{x,l}$.

Theorem 7.10. *We have the limit*

$$m_l = \lim_{x \to +\infty} E(I_{x,l}) = C_K(l)^{-1} - l.$$

Proof. Let us assume that $N_{x,l} = k$, then the average value of the function $I_{x,l}$ is

$$\frac{x - lk}{k+1}.$$

That is we have the conditional expectation

$$E(I_{x,l} \mid N_{x,l} = k) = \frac{x - lk}{k+1}$$

and from this the equality

$$E(I_{x,l}) = E\left(\frac{x - lN_{x,l}}{N_{x,l} + 1}\right) \tag{7.28}$$

follows. By Theorem 7.7 we have the convergence

$$\lim_{x \to +\infty} \frac{N_{x,l}}{x} = C_K(l) \quad \text{almost surely}$$

which implies

$$\lim_{x \to +\infty} \frac{x - lN_{x,l}}{N_{x,l} + 1} = C_K(l)^{-1} - l \quad \text{almost surely.} \tag{7.29}$$

For $l > 0$ we have $N_{x,l} \leq x/l$ from which it follows that the variables $(x - lN_{x,l})(N_{x,l} + 1)^{-1}$ are bounded for $0 \leq x < +\infty$ from this fact and (7.29), under consideration of (7.28), the assertion follows. For $l = 0$ the result is trivial since $I_{x,0}$ is uniformly distributed on $[0, 1]$.   □

Another proof can be given by starting from

$$m_l = \int_0^1 yG'(y)dy.$$

This method is used in the proof of the following theorem.

Theorem 7.11. *We have the equality*

$$\sigma^2 = \lim_{x \to \infty} V(I_{x,l})$$
$$= 1 - 4l + 2l^2 - \left(C_K(l)^{-1} - l\right)^2 - 2\frac{(1-l)^3}{3}\frac{F_2(l)}{C_K(l)}$$
$$+ 4\frac{1}{C_K(l)}\int_0^\infty \left\{\frac{(1-l)e^{-(1-l)t} - e^{-t}}{t} - \frac{e^{-t} - e^{-(1-l)t}}{t^2}\right\}\Psi(lt)dt.$$

Proof. If one denotes by $G'_l(h)$ the derivative with respect to h of $G_l(h)$ (which exists for $h \neq 1 - l$) then we have the following expression for the second moment:

$$\lim_{x \to \infty} E\left(I_{x,l}^2\right) = \int_0^1 h^2 G'_l(h)dh$$
$$= [h^2 G_l(h)]_0^1 - \int_0^1 2hG_l(h)dh$$
$$= 1 - \int_0^1 2hG_l(h)dh.$$

We thus have to integrate the second term by using the expression that we obtained in Theorem 7.1.

We then get

$$\int_0^1 h^2 G_l'(h)dh = 1 - \int_0^{1-l} 2hh \frac{F_2(l)}{C_K(l)}dh - \int_{1-l}^1 2h \left(2 - 2\frac{\int_0^\infty e^{-th}\Psi(lt)dt}{C_K(l)}\right) dh$$

$$= 1 - \frac{2(1-l)^3}{3}\frac{F_2(l)}{C_K(l)} - 2(2l - l^2)$$

$$+ 4\frac{1}{C_K(l)} \int_0^\infty \left[\int_{1-l}^1 he^{-th}dh\right] \Psi(lt)dt.$$

By successive integrations by part we get

$$\int_{1-l}^1 he^{-th}dh = \frac{(1-l)e^{-(1-l)t} - e^{-t}}{t} - \frac{e^{-t} - e^{-(1-l)t}}{t^2}$$

and the result follows by substitution and using the expression for the first moment in Theorem 7.10. □

In particular one gets $\sigma_1 = 0.28...$ [Bankövi (1962)]. Picture plots of the mean and variance are done in Figure 7.6.

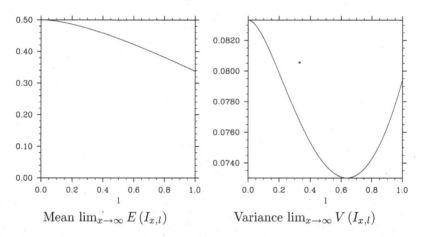

Mean $\lim_{x\to\infty} E\left(I_{x,l}\right)$ Variance $\lim_{x\to\infty} V\left(I_{x,l}\right)$

Fig. 7.6   Limit mean and variance of $I_{x,l}$ when $x \to \infty$ in terms of $l \in [0,1]$

Chapter 8

Parking cars with spin but no length

Following [Itoh and Shepp (1999)], we consider a variant of Rényi's random packing procedure where cars are allowed to park with some spin on a 1-dimensional street. Cars arrive to park on a block of length x, sequentially. Each car has, independently, spin up or spin down, with probability $0 < p \leq 1$, for spin up, and $q = 1 - p$, for spin down, respectively. Each car tries to park at a uniformly distributed random point $t \in [0, x]$. If t is within distance 1 of the location of a previously parked car of the same spin, or within distance a of the location of a previously parked car of the opposite spin then the new car leaves without parking and the next car arrives, until saturation.

In Section 8.1 a vector integral equations, which is similar to the integral equations for Rényi's problem is derived. In Section 8.2 a proof of the convergence of the limit packing density $c(p, a)$ is given using the integral equations. A few explicitly solvable cases are treated in Section 8.3 and some general solution methods are explained in Section 8.4. Another solution by power series is given in Section 8.5 which allows to solve explicitly one more case and give a rapidly convergent solution. This solution is computed in Section 8.6. Using it we show in Figures 8.2-8.4 that $c(p, a)$ is neither monotonic in a for fixed p, nor is it (see Figures 8.5-8.7) monotone in p for fixed a, in general.

Our model may be applied to RSA in the Widom-Rowlinson binary gas model [Widom and Rowlinson (1970); MacKenzie (1962); Van Lieshout (2006)] in which there are two kinds of particles, with different distance requirements depending on whether two particles are of the same type or different. Their model also assumes the distance between like species is shorter than the distance between unlike species, and we focus also on this case, but the other case is also interesting.

8.1   Integral equations

Cars arrive to park on a block of length x, sequentially. Each car has a spin up or down, with probability $0 < p \leq 1$, for spin up, independently, and each chooses to park at a uniformly distributed random point $t \in [0, x]$. If t is within distance 1 of the location of a previously parked car of the same spin, or within distance a of the location of a previously parked car of the opposite spin then the new car leaves without parking and the next car arrives, until saturation, when no more cars can park on $[0, x]$.

We give a formula for $c(p, a)$ usable for numerical calculation when $\frac{1}{2} \leq a \leq 1$ which will suffice to demonstrate the method. The cases $a < \frac{1}{2}$, and $a > 1$ can be treated by the same method, but the calculations are more tedious, and we have not carried them out. The expected number of cars for a block of length x can be written as an integral equation which in turn gives rise to non-commuting matrices, and a new technique is found to deal with these. In the special case, $p = 1/2$, the matrices do commute and the usual methods for commuting matrices, can be used to give a more conventionally explicit value for $c\left(\frac{1}{2}, a\right)$, which we give in Theorem 8.8.

The method could be extended (but we have not done it) for the case of many spin types, $i = 1, \ldots, n$ and with arbitrary probabilities, p_i, for cars of each spin type, i, and for an arbitrary matrix of allowed distances, $a_{i,j}$ between cars of spin i, j.

The case $a = 1$ is equivalent to Rényi's random packing problem considered in Chapter 3. The generalization of Rényi's methods to spin problem requires the introduction of more than one function and thus lead to vector recurrence and matrix equations. [Rényi (1958)] emphasizes the 1-dimensional character of the problem. In a sense, the spin problems are not 1-dimensional, but we show that the densities can be computed very accurately.

After the first car with a spin parks on a finite interval, we have car parking problems on two smaller intervals. Let $f_{\uparrow\uparrow}(x)$ denote the expected number of up-cars in saturation in a block of length x with two up-cars at the endpoints, and let $f_{\downarrow\downarrow}(x)$ and $f_{\uparrow\downarrow}(x)$ be defined similarly, again, keeping track only of up-cars. On the interval $[0, x]$ we can put up cars in $[1, x - 1]$ while we can put down cars in $[a, x - a]$ (see Figure 8.1). Thus the probabilities of having up-cars, respectively down cars are respectively

$$\frac{p(x-2)}{p(x-2) + q(x-2a)} \text{ and } \frac{q(x-2a)}{p(x-2) + q(x-2a)}.$$

We have the recursion relation

$$f_{\uparrow\uparrow}(x) = \frac{p}{x-2p-2qa} \int_1^{x-1} 1 + f_{\uparrow\uparrow}(u) + f_{\uparrow\uparrow}(x-u)du$$
$$+ \frac{q}{x-2p-2qa} \int_a^{x-a} f_{\uparrow\downarrow}(u) + f_{\uparrow\downarrow}(x-u)du.$$

By multiplying by $x - 2p - 2qa$ and passing to other side we get

$$x f_{\uparrow\uparrow}(x) = p \int_1^{x-1} 1 + f_{\uparrow\uparrow}(u) + f_{\uparrow\uparrow}(x-u)du$$
$$+ q \int_a^{x-a} f_{\uparrow\downarrow}(u) + f_{\uparrow\downarrow}(x-u)du + (2p + 2qa)f_{\uparrow\uparrow}(x).$$

Then dividing by x we get for $f_{\uparrow\uparrow}$ and similarly for $f_{\downarrow\downarrow}$ and $f_{\uparrow\downarrow}$ for $x > 2$:

$$f_{\uparrow\uparrow}(x) = p \int_1^{x-1} (f_{\uparrow\uparrow}(u) + f_{\uparrow\uparrow}(x-u) + 1)\frac{du}{x}$$
$$+ q \int_a^{x-a} (f_{\uparrow\downarrow}(u) + f_{\uparrow\downarrow}(x-u))\frac{du}{x} + \left(p\frac{2}{x} + q\frac{2a}{x}\right) f_{\uparrow\uparrow}(x)$$
$$f_{\downarrow\downarrow}(x) = p \int_a^{x-a} (f_{\uparrow\downarrow}(u) + f_{\uparrow\downarrow}(x-u) + 1)\frac{du}{x}$$
$$+ q \int_1^{x-1} (f_{\downarrow\downarrow}(u) + f_{\downarrow\downarrow}(x-u))\frac{du}{x} + \left(p\frac{2a}{x} + q\frac{2}{x}\right) f_{\downarrow\downarrow}(x)$$
$$f_{\uparrow\downarrow}(x) = p \int_1^{x-a} (f_{\uparrow\uparrow}(u) + f_{\uparrow\downarrow}(x-u) + 1)\frac{du}{x}$$
$$+ q \int_a^{x-1} (f_{\uparrow\downarrow}(u) + f_{\downarrow\downarrow}(x-u))\frac{du}{x}$$
$$+ \left(p\frac{1+a}{x} + q\frac{1+a}{x}\right) f_{\uparrow\downarrow}(x)$$

where we have used $f_{\uparrow\downarrow}(x) = f_{\downarrow\uparrow}(x)$ which holds by symmetry (Figure 8.1),

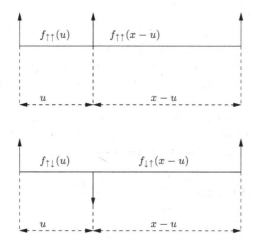

Fig. 8.1 Parking cars with spin in a block of length x with two up-cars at the end points

Note that for $0 < x < 2$ we can write $f_\sigma(x)$ explicitly for $1/2 < a < 1$ and $\sigma = \uparrow\uparrow, \downarrow\downarrow$ and $\uparrow\downarrow$, (this gets somewhat more complicated for other values of a). Indeed, for $x < 2$, and $1/2 \leq a \leq 1$, it is clear that

$$f_{\uparrow\uparrow}(x) = 0 \text{ if } 0 < x < 2,$$

$$f_{\downarrow\downarrow}(x) = \begin{cases} 0 & \text{if } 0 < x < 2a, \\ 1 & \text{if } 2a < x < 2, \end{cases}$$

$$f_{\uparrow\downarrow}(x) = \begin{cases} 0 & \text{if } 0 < x < 1 + a, \\ p & \text{if } 1 + a < x < 2. \end{cases}$$

8.2   Existence of the limit packing density

The convergence of the expected number to a constant of our problem is first shown in [Penrose (2001)] in mathematically rigorous way as an application of his law of large numbers for spatial point processes. Here we give a more elementary proof that uses the tools that we have introduced before. We use the methodology of Chapter 5 since here we are only looking after the existence of the limit itself, not its actual determination.

It follows easily from the integral equations that if the limits

$$c_\sigma = \lim_{x \to \infty} \frac{f_\sigma(x)}{x}$$

exist then all c_σ are equal (it suffices to substitute the asymptotic form in the integral equations to conclude). But it is false that one can erase the σ's from the integral equations and still get the right limit.

We have seen in Section 3.4 that Rényi's random cube packing density is best expressed as packing in the circle. For our problem, in a circle of length x after one puts the first car $\uparrow$ with probability p or $\downarrow$ with probability q, one is reduced to the case of $f_{\uparrow\uparrow}(x)$ or $\downarrow\downarrow$ and thus with the following expression for the packing density on the circle:

$$c(p, a, x) = \frac{p(f_{\uparrow\uparrow}(x) + 1) + q f_{\downarrow\downarrow}(x)}{x}.$$

It is reasonable to expect, based on Theorem 3.4 for Rényi's problem that this expression would converge very fast and numerical evidence seems to indicate that. But we are only able to prove the convergence of $c(p, a, x)$.

In Equation (8.1) the range of integral varies. This is significant for the value of the limit but not for the proof of their existence. The first basic fact is that $f_{\uparrow\uparrow}(x) \leq x$ and the same for $f_{\downarrow\downarrow}(x)$ and $f_{\uparrow\downarrow}(x)$.

Therefore we rewrite the equations in the following way:

$$\begin{cases} (x - 2a)f_{\uparrow\uparrow}(x) = 2p \int_a^{x-a} f_{\uparrow\uparrow}(u)du + 2q \int_a^{x-a} f_{\uparrow\downarrow}(u)du \\ \qquad\qquad + 2p \int_{x-1}^{x-a} (f_{\uparrow\uparrow}(x) - f_{\uparrow\uparrow}(u))du + p(x-2) \\ (x - 2a)f_{\downarrow\downarrow}(x) = 2p \int_a^{x-a} f_{\uparrow\downarrow}(u)du + 2q \int_a^{x-a} f_{\downarrow\downarrow}(u)du \\ \qquad\qquad + 2q \int_{x-1}^{x-a} (f_{\downarrow\downarrow}(x) - f_{\downarrow\downarrow}(u))du + p(x-2a) \\ (x - 2a)f_{\uparrow\downarrow}(x) = p \int_a^{x-a} f_{\uparrow\uparrow}(u)du + \int_a^{x-a} f_{\uparrow\downarrow}(u)du \\ \qquad\qquad + q \int_a^{x-a} f_{\downarrow\downarrow}(u)du \\ \qquad\qquad + \int_{x-1}^{x-a} (f_{\uparrow\downarrow}(x) - f_{\uparrow\downarrow}(u))du + p(x-1-a). \end{cases}$$

The main trouble comes from the term in the integrals on $[a, x - a]$, the remaining ones are comparatively negligible. Thus we can write the equations as

$$\begin{cases} f_{\uparrow\uparrow}(x) = \frac{1}{x-2a} \int_a^{x-a} 2p f_{\uparrow\uparrow}(u) + 2q f_{\uparrow\downarrow}(u)du + O(1) \\ f_{\downarrow\downarrow}(x) = \frac{1}{x-2a} \int_a^{x-a} 2p f_{\uparrow\downarrow}(u) + 2q f_{\downarrow\downarrow}(u)du + O(1) \\ f_{\uparrow\downarrow}(x) = \frac{1}{x-2a} \int_a^{x-a} p f_{\uparrow\uparrow}(u) + f_{\uparrow\downarrow}(u) + q f_{\downarrow\downarrow}(u)du + O(1). \end{cases} \tag{8.3}$$

For the vector $f = (f_{\uparrow\uparrow}, f_{\downarrow\downarrow}, f_{\uparrow\downarrow})$ it is apparent that the corresponding matrix is

$$\begin{pmatrix} 2p & 0 & 2q \\ 0 & 2q & 2p \\ p & q & 1 \end{pmatrix}$$

whose eigenvalues are 0, 1 and 2 and the corresponding left eigenvectors are

$$(1, 1, -2), (p, -q, q - p) \text{ and } (p^2, q^2, 2pq).$$

The corresponding eigenfunctions are

$$\begin{cases} f_0(x) = f_{\uparrow\uparrow}(x) + f_{\downarrow\downarrow}(x) - 2f_{\uparrow\downarrow}(x) \\ f_1(x) = p f_{\uparrow\uparrow}(x) - q f_{\downarrow\downarrow}(x) + (q - p)f_{\uparrow\downarrow}(x) \\ f_2(x) = p^2 f_{\uparrow\uparrow}(x) + q^2 f_{\downarrow\downarrow}(x) + 2pq f_{\uparrow\downarrow}(x) \end{cases}$$

and the rest of the proof will consider those eigenfunctions one by one.

Lemma 8.1. *We have $f_0(x) = O(1)$.*

Proof. This follows directly from Equation (8.3). □

Lemma 8.2. *We have $f_1(x) = O(\log x)$.*

Proof. By using Equation (8.3) we get the equation

$$f_1(x) = \frac{1}{x - 2a} \int_a^{x-a} f_1(t)dt + \phi(x)$$

with $|\phi(x)| \le C$ and C some constant. This is rewritten as

$$f_1(x) = \frac{1}{x} \int_a^{x-a} f_1(t)dt + \frac{2a}{x} f_1(x) + \frac{x-2a}{x} \phi(x)$$
$$= \frac{1}{x} \int_0^x f_1(t)dt + \phi_2(x)$$

with

$$\phi_2(x) = \frac{2a}{x} f_1(x) + \frac{x-2a}{x} \phi(x) - \frac{1}{x} \int_0^a f_1(t)dt - \frac{1}{x} \int_{x-a}^x f_1(t)dt.$$

We have $\phi_2(x) = O(1)$ since $f_1(x) = O(x)$. If one writes $F_1(x) = \int_0^x f_1(t)dt$ then the equation is

$$F_1'(x) = \frac{1}{x} F_1(x) + \phi_2(x)$$

and the solutions are

$$\begin{cases} F_1(x) = Cx + x \int_{x_0}^x \frac{\phi_2(t)}{t} dt \\ f_1(x) = C + \phi_2(x) + \int_{x_0}^x \frac{\phi_2(t)}{t} dt \end{cases}$$

with $x_0 \in \mathbb{R}_+^*$ and C some constant. The upper bound follows by a simple integration. $\square$

We note that the recursion of the kind satisfied by f_1 are part of the theory of one-sided random packing [Itoh and Mahmoud (2003)] where after a car is parked one parks only on the left (or right).

The function f_2 is more complicated to treat:

Lemma 8.3. *We have*

$$f_2(x) = c(p,a)x + O(1)$$

for some constant $c(p,a)$.

Proof. We have from Equation (8.2) and $x > 3$ the following equation

$$f_2(x) = \frac{2}{x - 2a} \int_a^{x-a} f_2(t)dt + b(x)$$

where

$$b(x) = \frac{2}{x-2a} \Big(p^3 \int_{x-1}^{x-a} f_{\uparrow\uparrow}(x) - f_{\uparrow\uparrow}(u)du$$
$$+ q^3 \int_{x-1}^{x-a} f_{\downarrow\downarrow}(x) - f_{\downarrow\downarrow}(u)du$$
$$+ pq \int_{x-1}^{x-a} f_{\uparrow\downarrow}(x) - f_{\uparrow\downarrow}(u)du + p(a-1) \Big).$$

Using the estimate $f_\sigma(x) = O(x)$ for $\sigma = \uparrow\uparrow, \downarrow\downarrow$ and $\uparrow\downarrow$ we have from Equations (8.2) the estimate $|f'_\sigma(x)| = O(1)$.

Using this bound we get the estimate

$$\left| \int_{x-1}^{x-a} f_\sigma(x) - f_\sigma(u) du \right| \leq \int_{x-1}^{x-a} \left| \int_u^x f'_\sigma(t) dt \right| du = O(1).$$

We thus finally get the estimation

$$b(x) = O\left(\frac{1}{x}\right).$$

We want to apply Theorem 5.1 in order to prove the required result. The estimate on $b(x)$ is what is needed but we also need to do a change of variable.

If we pose $g_2(x) = f_2(a + ax)$ and $b_2(x) = b(xa + 2a)$ we get

$$g_2(x+1) = \frac{2}{x} \int_0^x g_2(t) dt + b_2(x) \text{ with } b_2(x) = O\left(\frac{1}{x}\right).$$

We are now looking after applying Theorem 5.1. We write

$$p_j = \sup_{j \leq x \leq j+1} b_2(x) \text{ and } H_n = \sum_{j \geq n} \frac{p_j}{j}.$$

Using the estimate on b_2 we get $p_j = O\left(\frac{1}{j}\right)$ and $H_n = O\left(\frac{1}{n}\right)$. This implies that

$$R_j = \frac{2j+1}{j} p_{j+1} + \frac{2(j+1)(j+3)}{j} H_{j+2} = O(1).$$

Now we have to estimate the sum

$$S_n = \frac{2^n}{n!} \sum_{j=1}^n \frac{j!}{2^j}.$$

We pose $f(j) = \frac{2^j}{j!}$ and we have that $f(j)$ decreases for $j \geq 3$. Thus we get

$$S_n = f(n) \sum_{j=1}^n \frac{1}{f(j)}$$
$$= 1 + f(n) \sum_{j=1}^{n-1} \frac{1}{f(j)}$$
$$\leq 1 + f(n) \sum_{j=1}^{n-1} \frac{1}{f(n-1)}$$
$$\leq 1 + \sum_{j=1}^{n-1} \frac{2}{n}$$
$$\leq O(1).$$

Thus we get from Theorem 5.1 that there exists a constant λ such that

$$\sup_{n \leq x \leq n+1} |g_2(x) - \lambda x - \lambda| \leq \frac{2^n}{n!} \sup_{1 \leq x \leq 2} |g_2(x) - \lambda x - \lambda|$$
$$+ \frac{2^a}{n!} \sum_{j=1}^n \frac{j!}{2^j} R_j$$
$$\leq \frac{2^n}{n!} O(1) + S_n O(1)$$
$$\leq O(1).$$

This proves the sought estimation of f_2.   □

Theorem 8.1. *We have the limits*

$$\lim_{x\to\infty}\frac{f_{\uparrow\uparrow}(x)}{x}=\lim_{x\to\infty}\frac{f_{\downarrow\downarrow}(x)}{x}=\lim_{x\to\infty}\frac{f_{\uparrow\downarrow}(x)}{x}=c(p,a).$$

Proof. As a consequence of Lemmas 8.1, 8.2 and 8.3 we get

$$\lim_{x\to\infty}\frac{f_0(x)}{x}=\lim_{x\to\infty}\frac{f_1(x)}{x}=0 \text{ and } \lim_{x\to\infty}\frac{f_2(x)}{x}=c(p,a).$$

Since we can express f_σ as linear combination of the functions f_0, f_1 and f_2 the result follow.   □

8.3   Laplace transform and explicitly solvable cases

Now take Laplace transforms ϕ (with lower limit at $x=2$) of f defined for $\lambda>0$, and use the starting relations in $x<2$, above to calculate

$$\phi_\sigma(\lambda)=\int_2^\infty e^{-\lambda x}f_\sigma(x)dx$$

where σ can be any one of the pairs, $\uparrow\uparrow,\downarrow\downarrow,\uparrow\downarrow$.

One obtains, after a calculation, the column-vector equation,

$$\Phi'(\lambda)+A(\lambda)\Phi(\lambda)+g(\lambda)=0 \tag{8.4}$$

where $\Phi(\lambda)=(\phi_{\uparrow\uparrow}(\lambda),\phi_{\downarrow\downarrow}(\lambda),\phi_{\uparrow\downarrow}(\lambda))^T$ and where $A(u)$ is the matrix,

$$\begin{pmatrix} 2p+2aq+\frac{2pe^{-u}}{u} & 0 & \frac{2qe^{-au}}{u} \\ 0 & 2ap+2q+\frac{2qe^{-u}}{u} & \frac{2pe^{-au}}{u} \\ \frac{pe^{-au}}{u} & \frac{qe^{-au}}{u} & 1+a+\frac{e^{-u}}{u} \end{pmatrix}$$

and $g(\lambda)=(g_{\uparrow\uparrow}(\lambda),g_{\downarrow\downarrow}(\lambda),g_{\uparrow\downarrow}(\lambda))^T$, where

$$\begin{cases} g_{\uparrow\uparrow}(\lambda)=\frac{pe^{-2\lambda}}{\lambda^2}+\frac{2pqe^{-a\lambda}}{\lambda^2}(e^{-(1+a)\lambda}-e^{-2\lambda}) \\ g_{\downarrow\downarrow}(\lambda)=\frac{pe^{-2\lambda}}{\lambda^2}+\frac{2p(1-a)e^{-2\lambda}}{\lambda}+\frac{2p^2e^{-a\lambda}}{\lambda^2}(e^{-(1+a)\lambda}-e^{-2\lambda}) \\ \quad+\frac{2qe^{-\lambda}}{\lambda^2}(e^{-2a\lambda}-e^{-2\lambda}) \\ g_{\uparrow\downarrow}(\lambda)=\frac{pe^{-2\lambda}}{\lambda^2}+\frac{p(1-a)e^{-2\lambda}}{\lambda}+p\frac{e^{-\lambda}}{\lambda^2}(e^{-(1+a)\lambda}-e^{-2\lambda}) \\ \quad+\chi(a\ge\frac{2}{3})q\frac{e^{-a\lambda}}{\lambda^2}(e^{-2a\lambda}-e^{-2\lambda}) \\ \quad+\chi(a<\frac{2}{3})\frac{q}{\lambda}((2-3a)e^{-2\lambda}+\frac{(e^{-2\lambda}-e^{-(2+a)\lambda})}{\lambda}). \end{cases}$$

The fundamental difference with Rényi's formulation is that Equation (8.4) is a matrix equation and that solving first order linear differential equations with matrix coefficients is much harder than scalar ones.

Proposition 8.1. *[Itoh and Shepp (1999)]* *(i) For any a we have $c(p = 1, a) = c_R$.*

(ii) For any p we have $c(p, a = 1) = pc_R$.

Proof. (i) If $p = 1$ then only up spins are put and thus we simply have Rényi's model. If $a = 1$ then spin becomes independent of the packing and we simply have Rényi's packing model where p of the spins are up and q are down. □

The case $p = 0$ allows to find a discontinuity in behavior:

Theorem 8.2. *[Itoh and Shepp (1999)]* *(i) We have $c(p = 0, a) = 0$.*

(ii) We have

$$\lim_{p \to 0} c(p, a) = 2 \int_0^\infty (e^{-(2a-1)u} - e^{-u})e^{-2\int_0^u \frac{1-e^{-v}}{v} dv} du.$$

Proof. If $p = 0$ then no up-cars can occur and thus (i) holds trivially.

When p is small but positive, then, essentially, only down cars arrive until these become saturated, and then exactly one up car can park in those spaces between adjacent down cars when the gap is in $[2a, 2]$ because we have assumed $\frac{1}{2} \le a < 1$.

In Rényi's interval parking cars of length 1 are parked and the constraint is that the gap is smaller than 1. The result follows by application of formula (7.24) which follows from Theorem 7.1. □

Another proof method is to use Equation (8.4) and remark that if $p = 0$ then the equations decouple and $f_{\downarrow\downarrow}$ satisfies a scalar integral equation of Rényi's type.

Another remarkable explicit solution can be obtained for $p = 1/2$. The formula is given in Theorem 8.8 and takes five lines to give. The proof is based on the commutation of the matrices $A(u)$ and is part of a more general algebraic formalism for linear differential equations of degree 1.

8.4 General solution methods

In cases other than $p = 0$, $p = 1/2$, $p = 1$, or $a = 1$, explicit representation of $c(p, a)$ appears to be impossible in terms of elementary functions, although we succeed in giving an algorithm for determining $c(p, a)$ to arbitrary precision, and a partially explicit formula for $c(p, a)$.

This method does not extend to the case where $A(\lambda)$ is a matrix unless the matrices, $A(\lambda)$, all commute. If the matrices commute and they are diagonalizable then there exist a basis of eigenvectors (e_1, e_2, e_3) for all $A(\lambda)$ and consequently the problem reduces itself to 3 different scalar problems, i.e. equations of Rényi's type. One can easily see that the matrices $A(\lambda)$ commute with each other only if $p = \frac{1}{2}$.

The function $\Phi(\lambda)$ satisfies $\lim_{\lambda \to \infty} \Phi(\lambda) = 0$ thus $\Phi(\lambda)$ satisfies the differential equation (8.4) with a condition at infinity. We cannot apply standard existence theorems for differential equations to this setting and thus we will provide below an explicit solution. We should keep in mind that the general technique for proving existence of solutions of ordinary differential equations is to use fixed point Theorem 4.3.

Below, we will need to deal with estimation of matrices, and we need something that replace the absolute value. From a norm N on $\mathbb{R}^n$ we define a norm on the space of $n \times n$ matrices by

$$\|A\| = \sup_{X \in \mathbb{R}^n \, : \, \|X\| = 1} \|AX\|.$$

The major property of this norm is that $\|AB\| \leq \|A\| \times \|B\|$, that is it is an algebra norm from which we can estimate matrix products.

It is also an elementary fact that if A has an eigenvalue λ then $\|A\| \geq |\lambda|$. If one takes the Euclidean norm $\|.\|_2$ on $\mathbb{R}^n$ then one can prove that we have

$$\|A\|_2 = \sqrt{\max \lambda_i} \text{ with } \lambda_i \text{ eigenvalues of } A^T A.$$

That is the norm $\|.\|_2$ is almost optimal on the space of $n \times n$ matrices.

Theorem 8.3. *[Itoh and Shepp (1999)] If* $0 < p < 1$ *then Equation* (8.4) *admits the solution*

$$\Phi(\lambda) = \sum_{n=1}^{\infty} \int_{\lambda < u_1 < u_2 \ldots < u_n} A(u_1)A(u_2) \ldots A(u_{n-1})g(u_n)du_1 \ldots du_n. \quad (8.5)$$

Proof. Equation (8.4) is rewritten as

$$\Phi'(\lambda) = -A(\lambda)\Phi(\lambda) - g(\lambda)$$

or

$$\Phi(\lambda) = \int_{\lambda}^{\infty} A(u)\Phi(u) + g(u)du = K(\Phi)(\lambda).$$

So, if one starts from $\Phi_0 = 0$ and define $\Phi_{n+1} = K(\Phi_n)$ then one gets

$$\begin{cases} \Phi_1(\lambda) = \int_{\lambda}^{\infty} g(u_1)du_1, \\ \Phi_2(\lambda) = \int_{\lambda}^{\infty} A(u_1)\Phi(u_1) + g(u_1)du_1 \\ \qquad = \Phi_1(\lambda) + \int_{\lambda}^{\infty} A(u_1)\left\{ \int_{u_1}^{\infty} g(u_2)du_2 \right\} du_1 \\ \qquad = \Phi_1(\lambda) + \int_{\lambda < u_1 < u_2} A(u_1)g(u_2)du_1 du_2. \end{cases}$$

More generally one finds

$$\Phi_n(\lambda) = \Phi_{n-1}(\lambda) + H_n(\lambda)$$

with

$$H_n(\lambda) = \int_{\lambda < u_1 < \cdots < u_n} A(u_1) \ldots A(u_{n-1}) g(u_n) du_1 \ldots du_n.$$

So, clearly Φ_n is the n-th partial sum of the series (8.5). We need to prove its convergence. First notice that g is integrable on $[a, \infty)$ for any $a > 0$ because g decreases exponentially, while the function A is bounded on that same interval. Thus we have for any algebra norm $\|.\|$ on the 3×3 matrices with $C = \sup_{\lambda \in [a,\infty)} \|A(\lambda)\|$ the following inequality:

$$\|H_n(\lambda)\| \leq \int_{\lambda < u_1 < \cdots < u_n} \|A(u_1)\| \ldots \|A(u_{n-1})\| g(u_n) du_1 \ldots du_{n-1} du_n$$
$$\leq C^{n-1} \int_{u_n=\lambda}^{\infty} \left\{ \int_{\lambda < u_1 < \cdots < u_n} du_1 \ldots du_{n-1} \right\} g(u_n) du_n.$$
$$(8.6)$$

Denote by $X_n(u_n)$ the inner integral. It can be evaluated directly to

$$X_n(u_n) = \frac{(u_n - \lambda)^{n-1}}{(n-1)!}.$$

This integral can also be interpreted as the volume of a simplex. Thus the n-th term of the series is bounded by

$$\int_\lambda^\infty C^{n-1} \frac{(u_n - \lambda)^{n-1}}{(n-1)!} g(u_n) du_n \qquad (8.7)$$

and this integral converges because of the exponential decrease of g.

But to prove the convergence of the series we have to bound further $H_n(\lambda)$. First one can remark that $g(u_n)$ is estimated by $\frac{1}{\lambda} e^{-2\lambda}$. Secondly we have to estimate the norm of the matrices $A(\lambda)$. We have the following limit

$$\lim_{\lambda \to \infty} A(\lambda) = A_\infty = \begin{pmatrix} 2p + 2aq & 0 & 0 \\ 0 & 2ap + 2q & 0 \\ 0 & 0 & 1 + a \end{pmatrix}. \qquad (8.8)$$

For the norm $\|.\|_2$ we have the estimation

$$\|A_\infty\|_2 = \max(2p + 2aq, 2ap + 2q, 1 + a) = C_{eig} < 2. \qquad (8.9)$$

So, we can find $C < 2$ and λ_0 such that for $\lambda > \lambda_0$ we have $\|A(\lambda)\|_2 < C$. Such an estimate will not be valid for small values of λ since

$$\lim_{\lambda \to 0} \|A(\lambda)\|_2 = \infty.$$

But in the estimation of Equation (8.6) it is the large values that matter. Thus we can estimate the term of the series in Equation (8.7) in the following way

$$
\begin{aligned}
C^{n-1}\int_\lambda^\infty \frac{(u_n-\lambda)^{n-1}}{(n-1)!}\frac{1}{u_n}e^{-2u_n}\,du_n &\leq C^{n-1}\int_\lambda^\infty \frac{(u_n-\lambda)^{n-2}}{(n-1)!}e^{-2u_n}\,du_n \\
&\leq C^{n-1}\frac{e^{-2\lambda}}{(n-1)!}\int_0^\infty u_n^{n-2}e^{-2u_n}\,du_n \\
&\leq C^{n-1}\frac{e^{-2\lambda}}{(n-1)!}\frac{(n-1)!}{2^{n-1}} \\
&\leq \left(\frac{C}{2}\right)^{n-1}e^{-2\lambda}.
\end{aligned}
$$

Thus the series (8.5) converges. □

Theorem 8.4. *[Itoh and Shepp (1999)] If the matrices $A(\lambda)$ commute then we have*

$$
\Phi(\lambda) = \int_\lambda^\infty \exp\left(\int_\lambda^u A(v)dv\right)g(u)du,
$$

where exp *is the matrix exponential (see Definition 4.4).*

Proof. Let us define $a(u) = \int_\lambda^u A(v)dv$. The n-th term of the series (8.5) is

$$
\begin{aligned}
H_n(\lambda) &= \int_{\lambda<u_1<\cdots<u_n} A(u_1)A(u_2)\ldots A(u_{n-1})g(u_n)du_1\ldots du_n \\
&= \int_{u_n=\lambda}^\infty \left\{\int_{\lambda<u_1<\cdots<u_n} A(u_1)\ldots A(u_{n-1})du_1\ldots du_{n-1}\right\}g(u_n)du_n \\
&= \int_\lambda^\infty \left\{\int_{\lambda<u_2<\cdots<u_n} a(u_2)A(u_2)\ldots A(u_{n-1})du_2\ldots du_{n-1}\right\}du_n \\
&= \int_\lambda^\infty \left\{\int_{\lambda<u_3<\cdots<u_n} \frac{a(u_3)^2}{2!}A(u_3)\ldots A(u_{n-1})du_3\ldots du_{n-1}\right\}du_n \\
&= \int_\lambda^\infty \frac{a(u_n)^{n-1}}{(n-1)!}g(u_n)du_n.
\end{aligned}
$$

The result follows by definition of the matrix exponential. □

It is easy to verify that, when $p = 1/2$,

$$
A(u) = \left(1+a+\frac{e^{-u}}{u}\right)I + \frac{e^{-au}}{u}B
$$

where B does not depend on u. Hence the matrices $A(u)$ commute. This allows us to carry out the explicit solution for Φ above. We get,

$$
\Phi(\lambda) = \int_\lambda^\infty \exp\left\{\int_\lambda^u A(v)dv\right\}g(u)du
$$

and it is a simple calculation, representing $g(u)$ as a linear combination of the (common) eigenvectors of $A(u)$ to obtain the constant. The answer is easiest to give using the methods of the non-commutative case, however, as we do at the end of the next section.

8.5 The power series solution

We proceed instead, for $p \neq 1/2$ to seek a 3×3-matrix, $\Psi(\lambda)$ satisfying

$$\Psi'(\lambda) = \Psi(\lambda)B(\lambda) \tag{8.10}$$

where

$$B(\lambda) = A(\lambda) - \left(2 + \frac{2}{\lambda}\right)I_3.$$

Theorem 8.5. *[Itoh and Shepp (1999)] The function $\Psi(\lambda)$ is a solution of the equation*

$$e^{2\lambda}\lambda^2\Psi(\lambda)\Phi(\lambda) = \int_\lambda^\infty \Psi(u)h(u)du \tag{8.11}$$

where $h(u) = u^2e^{2u}g(u)$.

Proof. Existence theory for differential equations guarantees the existence of Ψ. Thus we can write

$$
\begin{aligned}
(\lambda^2 e^{2\lambda}\Psi(\lambda)\Phi(\lambda))' &= \lambda^2 e^{2\lambda}\left\{(\tfrac{2}{\lambda}+2)\Psi(\lambda)\Phi(\lambda) + \Psi'(\lambda)\Phi(\lambda) + \Psi(\lambda)\Phi'(\lambda)\right\} \\
&= \lambda^2 e^{2\lambda}\left\{(\tfrac{2}{\lambda}+2)\,\Psi(\lambda)\Phi(\lambda) + \Psi(\lambda)B(\lambda)\Phi(\lambda)\right. \\
&\quad \left. + \Psi(\lambda)(-A(\lambda)\Phi(\lambda) - g(\lambda))\right\} \\
&= -\Psi(\lambda)\lambda^2 e^{2\lambda}g(\lambda)
\end{aligned}
$$

where we have used Equation (8.4). Now in order to conclude, we need to prove the limit:

$$\lim_{\lambda\to\infty} e^{2\lambda}\lambda^2\Psi(\lambda)\Phi(\lambda) = 0. \tag{8.12}$$

From the proof of Theorem 8.3 it is clear that $\Phi(\lambda) = O\left(e^{-2\lambda}\right)$. Let us write $\Psi_b(\lambda) = e^{2\lambda}\Psi(\lambda)$. We have

$$\Psi'_b(\lambda) = \Psi_b(\lambda)\left[B(\lambda) + 2I_3\right]$$

and

$$B(\lambda) + 2I_3 = A(\lambda) - \frac{2}{\lambda}I_3.$$

By Equations (8.8) and (8.9) the eigenvalues of A_∞ are strictly lower than 2. Thus there exist $\alpha < 2$ and $C > 0$ such that

$$\|\Psi_b(\lambda)\|_2 \leq Ce^{\alpha\lambda}.$$

This implies the limit (8.12), the convergence of $\int_\lambda^\infty \Psi(u)h(u)du$ and Equality (8.11). $\qquad\square$

Note that we have the following expression for the function $h(u) = (h_{\uparrow\uparrow}(u), h_{\downarrow\downarrow}(u), h_{\uparrow\downarrow}(u))$:

$$\begin{cases} h_{\uparrow\uparrow}(u) = p + 2pq(e^{-(2a-1)u} - e^{-au}) \\ h_{\downarrow\downarrow}(u) = p + 2p(1-a)u + 2p^2(e^{-(2a-1)u} - e^{-au}) + 2q(e^{-(2a-1)u} - e^{-u}) \\ h_{\uparrow\downarrow}(u) = p + p(1-a)u + p(e^{-au} - e^{-u}) + \chi(a \geq \frac{2}{3})q(e^{-(3a-2)u} - e^{-au}) \\ \qquad + \chi(a < \frac{2}{3})q((2-3a)u + 1 - e^{-au}). \end{cases}$$

Now we would like to write a solution for Ψ that would enable us to calculate the limit $c(p, a)$. We have not yet specified the initial condition on Ψ. One might choose $\Psi(0) = I$ so that we have a formula for $\lambda^2 \Phi(\lambda)$, but this is not a good idea because then $\Psi(\lambda)$ will not be analytic at $\lambda = 0$. Instead, we will choose $\Psi(0)$ to be a degenerate matrix of 3 identical rows.

Theorem 8.6. *[Itoh and Shepp (1999)] There exists a solution $\Psi(\lambda)$ of Equation (8.10) which is analytic at 0 with Ψ having three identical rows.*

Proof. Let us write the power series solution as

$$\Psi(\lambda) = \sum_{n=0}^{\infty} \Psi_n \lambda^n. \qquad (8.13)$$

By expanding in power series Equation (8.10) we get

$$\sum_{n>0} n\Psi_n \lambda^{n-1} = \sum_{n\geq0} \Psi_n \lambda^n \sum_{n\geq-1} B_n \lambda^n \qquad (8.14)$$

with

$$B_{-1} = \begin{pmatrix} -2q & 0 & 2q \\ 0 & -2p & 2p \\ p & q & -1 \end{pmatrix}, \; B_0 = \begin{pmatrix} 2aq - 2 & 0 & -2aq \\ 0 & 2ap - 2 & -2ap \\ -ap & -aq & -1 \end{pmatrix}$$

and $B_k = \begin{pmatrix} 2p\frac{(-1)^{k+1}}{(k+1)!} & 0 & -2q\frac{(-a)^{k+1}}{(k+1)!} \\ 0 & -2p\frac{(-1)^{k+1}}{(k+1)!} & 2p\frac{(-a)^{k+1}}{(k+1)!} \\ p\frac{(-a)^{k+1}}{(k+1)!} & q\frac{(-a)^{k+1}}{(k+1)!} & \frac{(-1)^{k+1}}{(k+1)!} \end{pmatrix}$ for $k \geq 1$.

The matrix B_{-1} has eigenvalues $0, -1$ and -2.

Equation (8.14) gives $\Psi_0 B_{-1} = 0$. Thus every line of Ψ_0 has to be an eigenvector of B_{-1} that is it has to be a multiple of $w_0 = (p^2, q^2, 2pq)$. Since we want the matrix to have identical rows, we are thus led to choosing

$$\Psi_0 = \begin{pmatrix} p^2 & q^2 & 2pq \\ p^2 & q^2 & 2pq \\ p^2 & q^2 & 2pq \end{pmatrix}.$$

For $n \geq 0$, the following equation follows from (8.14) by expansion:

$$\Psi_{n+1}((n+1)I - B_{-1}) = \sum_{k=0}^{n} \Psi_{n-k}B_k$$

or

$$\Psi_{n+1} = \left(\sum_{k=0}^{n} \Psi_{n-k}B_k \right) ((n+1)I - B_{-1})^{-1}.$$

This specifies Ψ_n for all $n \geq 0$, recursively, because the matrix $(n+1)I - B_{-1}$ is never singular since its eigenvalues are $n+1, n+2, n+3$ and hence non-zero.

A matrix M has all rows identical if and only if $(1, -1, 0)M = (0, 1, -1)M = 0$. So, clearly for all n the matrix Ψ_n has identical rows. We thus write

$$\Psi(\lambda) = \begin{pmatrix} \psi(\lambda) \\ \psi(\lambda) \\ \psi(\lambda) \end{pmatrix} \text{ and } \Psi_n = \begin{pmatrix} \psi_n \\ \psi_n \\ \psi_n \end{pmatrix}.$$

The vector valued analytic function $\psi(\lambda)$ therefore satisfies

$$\psi'(\lambda) = \psi(\lambda)B(\lambda)$$

with $B(\lambda)$ having a singularity at 0. The singularity is regular (see Definition 4.3) thus there exists a matrix-valued function $z \mapsto C(z)$ such that $C(z)z^{B-1}$ is solution of the equation with $C(0) = I_3$. Such solutions are obtained by power series (see Theorem 4.7) thus the series (8.13) converges and is thus a solution of the equation. $\square$

Theorem 8.7. *[Itoh and Shepp (1999)] We have the expression*

$$c(p, a) = \int_0^{\infty} \langle \psi(u), h(u) \rangle du. \tag{8.13}$$

Proof. By Theorem 8.1 we know that

$$(c(p, a), c(p, a), c(p, a))^T = \lim_{x \to \infty} \frac{1}{x}(f_{\uparrow\uparrow}(x), f_{\downarrow\downarrow}(x), f_{\uparrow\downarrow}(x))^T.$$

By the Tauberian Theorem 3.2 we know that

$$(c(p, a), c(p, a), c(p, a))^T = \lim_{\lambda \to 0} \lambda^2 \Phi(\lambda).$$

Using Equation (8.11) we get

$$\int_0^{\infty} \Psi(u)h(u)du = (c(p, a), c(p, a), c(p, a))^T$$

which simplifies to

$$c(p, a) = \int_0^\infty \langle \psi(u), h(u) \rangle du,$$

that is the required result.   □

Theorem 8.8. *[Itoh and Shepp (1999)] If $p = 1/2$ then we have*

$$c(1/2, a) = \int_0^\infty e^{(a-1)u - \int_0^u (\frac{1-e^{-v}}{v} + \frac{1-e^{-av}}{v}) dv} \frac{h_{\uparrow\uparrow}(u) + h_{\downarrow\downarrow}(u) + 2h_{\uparrow\downarrow}(u)}{4} du$$

$$(8.14)$$

where $h(u)$ is given explicitly above, for $p = 1/2$, by

$$\begin{cases} h_{\uparrow\uparrow}(u) = \frac{1}{2} + \frac{1}{2}(e^{-(2a-1)u} - e^{-au}) \\ h_{\downarrow\downarrow}(u) = \frac{1}{2} + (1-a)u + \frac{1}{2}(-(2a-1)u - e^{-au}) + (e^{-(2a-1)} - e^{-u}) \\ h_{\uparrow\downarrow}(u) = \frac{1}{2} + \frac{1}{2}(1-a)u + \frac{1}{2}(e^{-au} - e^{-u}) + \chi(a \geq \frac{2}{3})\frac{1}{2}(e^{-(3a-2)u} - e^{-au}) \\ \qquad + \chi(a < \frac{2}{3})\frac{1}{2}((2-3a)u + 1 - e^{-au}). \end{cases}$$

Proof. If $p = 1/2$ then all the matrices $A(\lambda)$ commute. Therefore the eigenvector $w_0 = (p^2, q^2, 2pq) = (1/4, 1/4, 1/2)$ is eigenvector for all $B(\lambda)$ and so for all B_k. Therefore we can write $\psi(u) = w_0 f(u)$ and Equation (8.10) simplifies to

$$f'(u) = f(u) \left(a - 1 + \frac{e^{-u} + e^{-au} - 2}{u} \right).$$

By solving the equation one finds

$$f(u) = \exp \left\{ (a-1)u - \int_0^u \frac{1 - e^{-v}}{v} + \frac{1 - e^{-av}}{v} dv \right\}$$

which directly implies formula (8.14).   □

Remark 8.1. In the general non-commutative case one could, in principle, interchange the sum and integral and integrate, term-by-term, to obtain,

$$c = \sum_{n \geq 0} \Psi_n \int_0^\infty u^n h(u) du$$

but this sum does not converge even in Rényi's case. To see this, note that this would mean in Rényi's case where $g(u) = \frac{e^{-u}}{u^2}$, that

$$c = \sum_{n \geq 0} \Psi_n n! = \int_0^\infty \Psi(u) e^{-u} du$$

where $\Psi_0 = 1$, and for $n \geq 0$,

$$(n+1)\Psi_n = \sum_{k=0}^n \Psi_{n-k} b_k.$$

But if the above series converged, then the function, Ψ, would be of exponential type 1, since $\Psi_n n!$ must tend to zero. But in Rényi's case we know that $\Psi(u) = e^{u - 2 \int_0^u \frac{1-e^{-v}}{v} dv} du$ and this is not entire of type 1 as we see by letting u large negative. However, we can still use the integral for numerical evaluation of $c(p, a)$. The formula obtained gives an algorithm, at least in principle, to find it.

8.6 Numerical computations

For Rényi's problem one has the following two expressions for the limit packing density:

$$C_R = \int_0^\infty \exp\left\{-2 \int_0^t \frac{1-e^{-u}}{u} du\right\} dt$$
$$= 2 \int_0^\infty e^{-u} \exp\left\{-2 \int_0^t \frac{1-e^{-u}}{u} du\right\} dt.$$

The second expression is much more practical for effective integration due to its exponential decrease at ∞. We consider how one can efficiently compute the number $c(p, a)$ in the general case.

A similar technique works for the general case, starting from the main formula (8.13), and integrating by parts we get,

$$c(p, a) = - \int_0^\infty u(\Psi(u)h(u))' du$$
$$= - \int_0^\infty u(\Psi(u)A(u) - \left(2 + \frac{2}{u}\right)I) h(u) + \Psi(u)h'(u)) du.$$

Now in making a numerical quadrature of the integral to obtain $c(p, a)$, one could calculate $\Psi(u)$ for each needed value of u either from the power series representation given in Section 8.5, or, alternatively, directly from the differential equation. If one used the former method, then if the step size for the quadrature is δ, then the quadrature sum, based on the trapezoid rule, which we will call $J(\delta)$ should have a power series with only even powers of δ, namely $J(\delta) = J(0) + c_2 \delta^2 + \ldots$, where $J(0)$ is the actual value of the integral, i.e. $c(p, a)$. If, instead, we use the differential equation to update the needed values of $\Psi(u)$, then, because of the inherent error of computing $\Psi(u_{n+1})$ from the value at u_n, the quadrature sum will have all powers of δ, and a smaller step size must be used when using Simpson's or Romberg's rule. Nevertheless, for ease of programming we choose the second method which seems to work quite well. A C program was written (available from [Itoh and Shepp (1999)]) for computing the value of $c(p, a)$ for all values of $p \in [0, 1]$ and $a \in [\frac{1}{2}, 1]$ to 5 decimal place accuracy, as was checked by halving the step size and verifying that the results agreed to at least 5 digit accuracy. If the step size is doubled, from that indicated (.01), then the last decimal place accuracy is lost.

We made direct simulations to make sure the numerical values. In Rényi's case, for the expected number $M(x)$ of cars in a street of length x, the form $\frac{M(x)+1}{x+1}$ gives a very good approximation of the parking constant c_R even for small x. By making numerical studies for our problem for the discrete street, it was found by experimental method that the form

$$c(p, a, x) = \frac{p^2 f_{\uparrow\uparrow}(x) + q^2 f_{\downarrow\downarrow}(x) + 2pq f_{\uparrow\downarrow}(x) + p}{x}$$

gives a good approximation to the $c(p, a)$ even for small x. We make use of this form for our simulation of seven decimal precision for the continuous street. We make 10,000 trials for each of the three f, $f_{\uparrow\uparrow}(x)$, $f_{\downarrow\downarrow}(x)$, $f_{\uparrow\downarrow}(x)$ and obtained $c(p, a, 15)$ and $c(p, a, 20)$ for every 0.05 of $1/2 \leq a \leq 1$ for $p = 0.5, 0.7, 0.95$. The values by our simulations agree with the values obtained by our numerical method up to three decimals.

Figures 8.2-8.4 plot $c(p, a)$ for fixed $p = .5, .7, .9$ as functions of $.5 < a < 1$ in steps of .01 (we have only considered this range of a). Figures 8.5-8.7 plot $c(p, a)$ for fixed $a = .5, .52, .54$ as functions of $0 < p < 1$. Note that $c(p, a)$ is not monotonic in p in Figures 8.5, 8.6, which at first was surprising to us. A post-facto and intuitive explanation is that for smaller p there will be, as parking nears saturation, more intervals with length, $1 < L < 2$ with down spins at each endpoint, than there would be such intervals with up spins at the endpoints, for larger p. In either case, the spin of the last car to park in such an interval is determined and must be the opposite spin to those at the ends. Note also that $c(p, a)$ is neither monotonic in a for fixed p in Figure 8.4. Here the explanation is even more subtle and post-facto. We believe it is that for fixed large p, close to 1, an interval, waiting for its last car and of length slightly larger than 2 is likely to have up-spin cars at its endpoints, and then is more likely to be filled by a car with down spin if a is smaller than when a is larger. Thus making a smaller keeps down the number of up-spin cars parking in the last available space. Finally, note the very strange oscillations in $c(p, a)$ in Figures 8.6 and 8.7 for $a = .52, .54$, near $p = 0$. An expansion of this region is given in Figure 8.8 for $a = .52$. Figure 8.9 shows $c(p, .6)$, again monotonic in p.

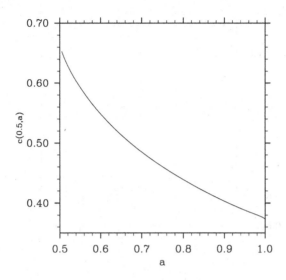

Fig. 8.2   Plot of $c(p,a)$ for fixed $p = .5$ as functions of $.5 < a < 1$

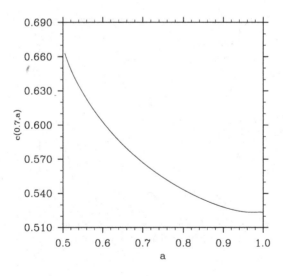

Fig. 8.3   Plot of $c(p,a)$ for fixed $p = .7$ as functions of $.5 < a < 1$

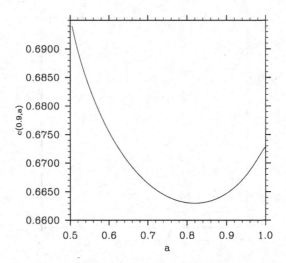

Fig. 8.4   Plot $c(p, a)$ for fixed $p = .9$ as functions of $.5 < a < 1$

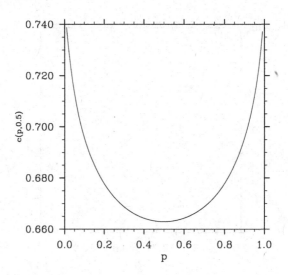

Fig. 8.5   Plot of $c(p, a)$ for fixed $a = .5$ as functions of $0 < p < 1$

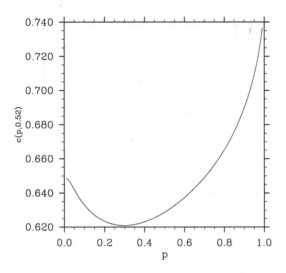

Fig. 8.6   Plot of $c(p, a)$ for fixed $a = .52$ as functions of $0 < p < 1$

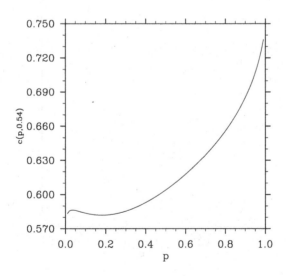

Fig. 8.7   Plot of $c(p, a)$ for fixed $a = .54$ as functions of $0 < p < 1$

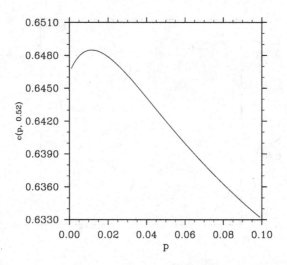

Fig. 8.8   An expansion of the region of the oscillations in Figure 8.6 is given

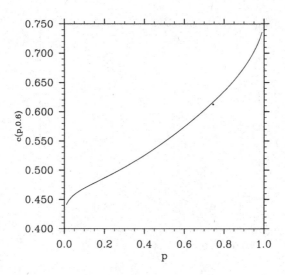

Fig. 8.9 $c(p, .6)$, is again monotonic in p

Chapter 9

Random sequential packing simulations

Due to the difficult nature of packing problems and of sequential random packing problems there is a need for computer simulations which although they cannot solve problems, that is giving rigorous proofs, can indicate what their solution, if any, should be.

We consider here general algorithms for sequential packing in a metric space. We relate the termination problem of those algorithms to the covering problem, which we describe in this chapter. Then we consider successively the problem of random sequential packing for spheres, cubes and cross polytope (unit ball for the Hamming distance) and the results obtained by simulations.

9.1 Sequential random packing and the covering problem

Perhaps the most general way of defining the notion of packing is to consider a set $\mathcal{E}$ in a metric space X.

Definition 9.1. Let X be a metric space, $\mathcal{E}$ a set of points of X then:
(i) For a $r > 0$, $\mathcal{E}$ is called a *r-packing* if for every $x \neq y \in \mathcal{E}$ we have

$$d(x,y) \geq r. \tag{9.1}$$

(ii) Given $r > 0$, an r-packing $\mathcal{E}$ is called *non-extensible* if for any r-packing $\mathcal{E}'$ such that $\mathcal{E} \subset \mathcal{E}'$ we have $\mathcal{E} = \mathcal{E}'$.
(iii) For a set $\mathcal{E}$ the packing radius is the largest r such that $\mathcal{E}$ is a r-packing.

In order to test if a set of points $\mathcal{E}$ defines a r-packing, we simply have to test that the inequalities (9.1) hold, that is we simply have to be able to compute the pairwise distances. It can be difficult in some cases, but in general this is reasonably easy.

If X is a metric space, $x \in X$ and $r > 0$ then we can define the open ball and closed ball:

$$B(x,r) = \{y \in X \ : \ d(x,y) < r\},$$
$$\overline{B}(x,r) = \{y \in X \ : \ d(x,y) \le r\}.$$

If $\mathcal{E}$ is an r-packing then the open balls $B(x, \frac{r}{2})$ are disjoint.

But non-extensibility is *a priori* much more difficult to test. In order to do this, we introduce the notion of covering:

Definition 9.2. Let X be a metric space, $\mathcal{E}$ a set of points of X and $r > 0$ then:

(i) $\mathcal{E}$ is called a *r-covering* if for every point x of X there exists a $z \in \mathcal{E}$ such that

$$d(x,z) \le r.$$

(ii) $\mathcal{E}$ is called a *strict-r-covering* if for every point x of X there exists a $z \in \mathcal{E}$ such that

$$d(x,z) < r.$$

Geometrically this is translated as follows: a set $\mathcal{E}$ is a r-covering, respectively strict r-covering if and only if the balls $\overline{B}(x,r)$, respectively $B(x,r)$ cover X completely.

We then have the following theorem:

Theorem 9.1. *Let $\mathcal{E}$ be an r-packing. The following are equivalent:*

(i) $\mathcal{E}$ is non-extensible.

(ii) $\mathcal{E}$ is a strict-r-covering.

Proof. Assume $\mathcal{E}$ is non-extensible. This is equivalent to saying that for any $z \in X$, $\mathcal{E} \cup \{z\}$ is not a packing that is there exists $x \in \mathcal{E}$ such that $d(x,z) < r$, i.e. that $\mathcal{E}$ is a strict-r-covering. $\square$

The covering problem asks for the minimal number of balls needed to cover the space. The covering problem is a kind of dual to the packing problem, which asks for the maximum number of non-overlapping balls that can be put into a space. The covering problem requires more complicated methods to be studied than the packing problem. Also, in some specific studies [Cohen, Honkala, Litsyn and Lobstein (1997); Schürmann (2009)] it appears that there is much less remarkable structures than for the packing problem.

A metric space X is called a finite packing metric space if for every $r > 0$ there exist $N(r)$ such that for any r-packing $\mathcal{E}$ the set $\mathcal{E}$ is finite and has at most $N(r)$ elements. Typically a finite packing metric space is a finite space (like, for example, $\{0,1\}^n$) or a compact space like the sphere, or a box, but not the whole Euclidean space.

Provided that we have a probability measure on X, we can define the notion of sequential random packing on a finite packing metric space:

Input: Finite packing metric space X and distance $r > 0$
Output: A non-extensible r-packing $\mathcal{E}$

$\mathcal{E} \leftarrow \emptyset$
while $\mathcal{E}$ is non-extensible **do**
    $\mathcal{F} \leftarrow$ set of $x \in X$ such that $\mathcal{E} \cup \{x\}$ is a r-packing
    $z \leftarrow$ a point of $\mathcal{F}$ chosen at random
    $\mathcal{E} \leftarrow \mathcal{E} \cup \{z\}$
end while

The two key difficulties for a given packing $\mathcal{E}$ are thus:

(i) Determine whether or not $\mathcal{E}$ is extensible.
(ii) Find a point z to extend $\mathcal{E}$ at random.

Of course, the solution of (ii) implies the one of (i). In general knowing that a packing is extensible does not imply that it is easy to find a point z at random. The point z is a proof that the packing is extensible. Thus it is named in computer science a *certificate*; that is it is easy to see that z extends a packing but finding such a z might be difficult.

To obtain a certificate is difficult but the following basic algorithm, if slow, allows to get one at random:

Input: Metric space X, $r > 0$ and an extensible r-packing $\mathcal{E}$
Output: A point $z \in X$ such that $\mathcal{E} \cup \{z\}$ is a r-packing

while
    $z \leftarrow$ a point of $\mathcal{E}$ chosen at random
    if $\mathcal{E} \cup \{z\}$ is a r-packing **do**
        return z
    end if
end while

The problem of this algorithm is that the while loop can last a very long time. Also we need to know *a priori* that $\mathcal{E}$ is extensible, which is not easy.

This class of algorithm is called *Las Vegas*: the running time is a random variable, but the results are always correct.

We will now introduce some of the notions that allow to solve the covering problem in some cases.

Definition 9.3. Let X be a metric space, $\mathcal{E}$ a set of points of X then:
(i) For $x \in \mathcal{E}$, the *Voronoi region* $V(\mathcal{E}, x)$ at x is:

$$V(\mathcal{E}, x) = \{z \in \mathcal{E} \text{ s.t. } d(z, y) > d(z, x) \text{ for all } y \neq x, y \in \mathcal{E}\}.$$

(ii) The *covering radius* of $x \in \mathcal{E}$ is

$$r_{cov}(\mathcal{E}, x) = \max_{y \in V(\mathcal{E}, x)} d(y, x).$$

(iii) The *covering radius* of $\mathcal{E}$ is

$$r_{cov}(\mathcal{E}) = \max_{x \in \mathcal{E}} r_{cov}(\mathcal{E}, x).$$

Obviously $\mathcal{E}$ is an r-covering if and only if $r_{cov}(\mathcal{E}) \leq r$. For strict-$r$-covering, we do not have such a general formulation but in practice, for the cases that we will consider, we will have $\mathcal{E}$ is a strict r-covering if and only if $r_{cov}(\mathcal{E}) < r$.

To find the Voronoi region for a distance d is a subject of active research. See for example [Chew (1998); Boissonat (1998)] for some algorithms that cover the L^1 and L^∞ cases. We will not need such generality in the cases that we consider. We will expose now the situation for the Euclidean case where the geometric picture is particularly elegant.

9.2 Random packing of spheres

For the Euclidean norm L^2 the conditions $\|x - c\| \leq \|x - c'\|$ in the definition of Voronoi regions define a family of half spaces. This implies that Voronoi regions are polytopes and an extensive theory has been developed in that case.

For a set $\mathcal{E} \subset \mathbb{R}^n$ a sphere $S(c, r)$ of center c and radius r is called an *empty sphere* if

(i) For all $x \in \mathcal{E}$ we have $\|x - c\| \geq c$.
(ii) The set $S(c, r) \cap \mathcal{E}$ is of affine rank $n + 1$.

A *Delaunay polytope* is defined as the convex hull of $S(c, r) \cap \mathcal{E}$ for $S(c, r)$ an empty sphere. If $\mathcal{E}$ is discrete then the set of Delaunay polytopes realize a packing of $\mathbb{R}^n$ but do not necessarily cover it completely. The centers of empty spheres are vertices of the Voronoi tessellation and more generally any k-dimensional face of the Delaunay tessellation is orthogonal to a $(n - k)$-dimensional face of the Voronoi tessellation and vice versa. The different structures are given in Figure 9.1. The Delaunay polytopes can be computed by a dual description computation, see for example [Dutour Sikirić, Schürmann and Vallentin (2009)] for an algorithm in a periodic setting. The covering radius is the maximum of the circumradius of all the empty spheres and thus can be computed easily once the Delaunay tessellation is known. Thus we can do efficient sequential random packing of spheres in Euclidean space.

The major problem in sphere packing theory is to find the sphere packing that maximize the packing density in Euclidean space, [Sloane (1984)]. The problem for lattices is resolved for dimension $n \leq 8$ and $n = 24$, see [Schürmann (2009)] for a recent presentation. For general packing the solution is known only in dimension 2 [Lagrange (1773); Thue (1910)] and 3 [Hales (2000, 2005)]. An asymptotic formula that provides an upper bound for the packing density, ρ_n, of n-dimensional spheres in an n-dimensional space as the space volume approaches ∞ is due to H. E. Daniels and reported in [Rogers (1958, 1964)]. It is

$$\sigma_n \sim \frac{(n+1)! e^{(n/2)-1}}{\sqrt{2}\Gamma(1+\frac{n}{2})(4n)^{n/2}}$$
$$\sim \frac{n}{e}\left(\frac{1}{\sqrt{2}}\right)^n.$$

In addition several investigators have looked into the random packing of 2-dimensional spheres. As reported in [Matheson (1974)] at first the technique were experimental but soon computer were used for simulations. Random sequential packing of spheres has been applied by [Bernal (1959)] to study the structure of liquids and since then it has been discussed by other authors, for example [Higuti (1960); Solomon (1967); Tanemura (1979)] and others.

Packing in dimension 3 are of interest in condensed matter theory as exemplified by [Lines (1979)]. In this kind of context one does not use sequential random packing but the notion of *compressed random packing*, i.e. when one pushes the wall of the packing box until one cannot improve the density any more. This led to an experimental density of 64% [Jaeger and Nagel (1992)] as opposed to a theoretical maximum of 74% for sphere packing. The theoretical basis of the notion of compressed random packing was put into question in [Torquato (2000)] where it was shown that it gives inconsistent result for practical and theoretical simulations. Instead the notion of *jammed random packing* was proposed: a packing is *jammed* if

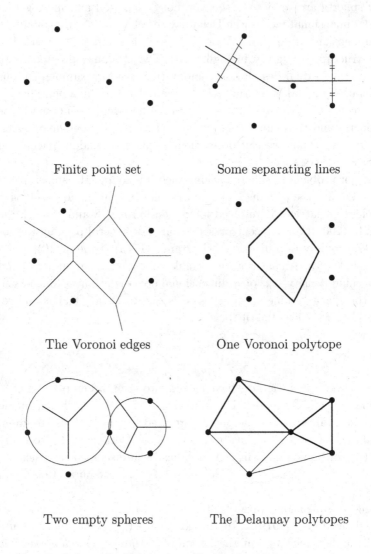

Fig. 9.1   Delaunay and Voronoi tessellations for a finite plane point set

no sphere can move while the other stay fixed. Using this notion a study [Song et al. (2008)] based on theoretical physics methods yielded that the random packing density could not be higher than 63.4%.

If the goal is to get good packing by random methods then a wholly different set of methods is required. In [Diaconis (2008)] a method based on

Monte Carlo Markov Chain is proposed where one does local improvement to the packing in order to get good ones. The method is studied but it is recognized there that due to the global nature of packing this class of methods based on local improvements does not work very well. Instead in [Donev, Torquato, Stillinger and Connelly (2004)] a method based on linear programming is proposed that can do global movement of the set of ball centers and thus improve globally the packing density even in jammed configurations.

9.3 Random packing of cubes

The Palásti model for random packing consider the random sequential packing of cubes $[0, 1]^n$ in the space $[0, x]^n$. The limit packing density is defined as

$$\beta_n = \lim_{x \to \infty} \frac{E(M_n(x))}{x^n}$$

with $M_n(x)$ the random variable of the number of cube $[0, 1]^n$ put in $[0, x]^n$. She conjectured [Palásti (1960)] that the constant β_n exists and that $\beta_n = \beta_1^n$.

This conjecture was extensively tested in [Blaisdell and Solomon (1970)] by computing β_2 and this seems to indicate that $\beta_2 \neq \beta_1^2$. The computational method was to consider packing of cubes of size $z + [0, 1]^n$ in $[0, x]^n$ for $z \in \frac{1}{a}\mathbb{Z}^n$ and $x \in \mathbb{Z}$ and using regression techniques to estimate the limit density β_2. The method is not rigorous for a number of reasons: the required power expansion of the packing density are not proved, the packing density itself is only estimated and the regression neglects higher order terms. But it is a reasonably efficient technique for statistical estimation. Furthermore as part of this work they estimated β_1 to be

$$\beta_1 = 0.747597920243398...$$

and they pointed out that their estimation was not based on the integral expression of $\beta_1 = C_R$ but on Dvoretzky Robbins theory (in Chapter 5). Other refutation of the Palásti's conjecture were done in [Akeda and Hori (1975, 1976)].

Computer estimation of constants such as β_n is very hard because the search space is very large and the constant itself is obtained as a limit. The dimension of the model is the driving difficulty. In [Blaisdell and Solomon (1982)] this was done for dimension 3 and 4 and the result are summarized

Table 9.1   Known estimation of Palásti's constant (from [Blaisdell and Solomon (1982)])

n	1	2	3	4
β_n	0.7476	0.5626	0.42623	0.32507
β_1^n	0.7476	0.5589	0.41783	0.31237

in Table 9.1. We do not know further work on the computation of this constant. Furthermore they proposed the following empirical formula, up to some higher order terms:

$$\beta_n^{1/n} - \beta_1 \simeq (n-1)(\beta_2^{1/2} - \beta_1) \qquad (9.2)$$

for $n = 3$, 4 which will be discussed in Chapter 12.

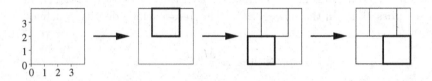

Fig. 9.2   The 2-dimensional case of the simplest sequential random cube packing process

In [Itoh and Ueda (1983)] a simple random cube packing model was proposed. In this model we consider an n-dimensional cube with side length 4 and a cubic lattice with unit side length. Cubes of side length 2 are put sequentially at random into the cube of side length 4 so that each vertex coincides with one of the lattice points. We continue until no place can be found in the large cube to place the smaller one. See in Figure 9.2 one illustration of this process. This model is simpler than Palásti's model for two reasons: we no longer have to take the limit of taking the containing cube to infinity and we are taking integral coordinate instead of real coordinates. Denote by γ_n the expectation of the packing density.

The packing density was initially estimated in [Itoh and Ueda (1983)] for some dimensions and the results were improved in [Itoh and Solomon (1986)], see Table 9.2. The results indicate that the quotients γ_{n+1}/γ_n seem to converge to 1. Furthermore the experimental formula (9.2) appears to fit the result with β_n replaced by γ_n for $n = 3$, 4 and 5. Thus it might be that β_{n+1}/β_n tends to 1 as the dimension increases.

A study on Table 9.2 indicates that

$$\gamma_n \sim \frac{1}{n^\alpha}$$

Table 9.2   Simple cubic random packing

n	γ_n	std. deviation	nr. trials	γ_{n+1}/γ_n
1	0.8348	0.2352	10000	0.8519
2	0.7112	0.2196	10000	0.8657
3	0.6157	0.1790	10000	0.8902
4	0.5481	0.1385	10000	0.8989
5	0.4927	0.1044	10000	0.9150
6	0.4508	0.0773	10000	0.9343
7	0.4212	0.0548	10000	0.9397
8	0.3958	0.0404	3000	0.9505
9	0.3762	0.0277	500	0.9652
10	0.3631	0.0178	100	0.9683
11	0.3516	0.0131	100	

for some empirical constant α as in the case of random sequential coding. A power law decay also seems reasonable for β_n instead of Palásti's conjecture.

In Figure 9.3 we give the repartition by number of cubes for the sequential random packing in the cube $[0,4]^n$ for $n = 3, 4$ and 5. We also give the information for the torus cube packing.

9.4   Random sequential coding by Hamming distance

Packing has applications in information theory as well as in physical and probabilistic models. More precisely suppose we have a $2r$-packing $\mathcal{E}$ in a metric space X. Then if $z \in X$ we can consider a point $x(z)$ that is the nearest to z in $\mathcal{E}$. If $d(x, z(e)) < r$ then the point $x(z)$ is uniquely defined and thus we have a way to code information with a packing. Note that the point $z(x)$ is not defined uniquely in general and that this *closest problem* is related to the covering problem.

One of the possible applications of spherical random packing is in recognition theory [Dolby and Solomon (1975)]. Suppose we have a set of objects $\mathcal{O}$ which we want to identify with known ones in a set $\mathcal{E}$. For every object in $\mathcal{O}$ we find the closest elements in $\mathcal{E}$. If this closest element is unique and if it is within distance r then we keep it, otherwise we reject it. [Dolby and Solomon (1975)] used Euclidean distance for such recognition problem of words and grades in school. Another consideration is in [Itoh and Hasegawa (1980)] to the problem of protein coding and its solution in nature by the DNA code.

However, the main application of such methods is in digital technology [Hamming (1947)]: a message consisting of a $\{0,1\}$ vector $x \in \{0,1\}^n$ is transmitted and along the way of transmission some errors could be

Table 9.3   Random packing density and
$n^{-\alpha}$

n	k	Packing density	$n^{-\alpha}$
3	2	0.43665	0.50333
4	2	0.38773	0.42052
5	2	0.34627	0.36578
6	2	0.31517	0.32640
7	2	0.28970	0.29642
8	2	0.26913	0.27269
9	2	0.25215	0.25334
10	2	0.23656	0.23720
11	2	0.22324	0.22349
12	2	0.21179	0.21166
13	2	0.20186	0.20133
14	2	0.19257	0.19222
15	2	0.18399	0.18411
16	2	0.17677	0.17683
17	2	0.17018	0.17026
4	3	0.12500	0.16059
5	3	0.12124	0.11964
6	3	0.09708	0.09406
7	3	0.07764	0.07675
8	3	0.06430	0.06435
9	3	0.05601	0.05509
10	3	0.04833	0.04794
11	3	0.04263	0.04228
12	3	0.03799	0.03769
13	3	0.03410	0.03392
14	3	0.03076	0.03076
15	3	0.02802	0.02808
16	3	0.02557	0.02579
17	3	0.02345	0.02381

introduced in x. Those errors are unavoidable due to the technological choices and we need a way to correct them. The idea is to associate to x a vector $x' = \phi(x) \in \{0,1\}^{n'}$ with $n' \geq n$ such that errors in the transmission of $\phi(x)$ can be recovered.

If the image by ϕ of $\{0,1\}^n$ is an r-packing in $\{0,1\}^{n'}$ and the function ϕ is injective then it is possible in principle to recover some errors. If $r = 2e + 1$ then if the transmission of x' has at most e errors then we can correct them. If $r = 2e$ then transmission with less than e errors can be corrected. If the number of errors is e then we may be in a situation where we cannot correct errors. Of course if the number of errors is greater than e then we cannot do anything, but at least the method allows some correction of errors and is the basis of today's electronic digital infrastructure.

Thus there is a need of finding packing in order to be able to transmit information efficiently. A (n, M, r) code is an r-packing in $\{0,1\}^n$ for the Hamming distance $d_1(x, x') = \sum_{i=1}^{n} |x_i - x'_i|$ with M codewords. For a fixed n and r, the higher M is the better is the code. We will give the result of some simulations of sequential random packing for a fixed n and r. Random strategies were actually used to give non-constructive proof of existence of some good packings in the pioneering works in information theory [Shannon (1948)]. But before that we will explain some of the obtained results for best packing in the hypercube.

Let us denote by $\delta_{n,r}$ the highest density of a non-extensible r-packing for the Hamming distance in $\{0,1\}^n$.

Theorem 9.2. *We have*

$$\frac{1}{\sum_{i=0}^{2e} \binom{n}{i}} < \delta_{n,r} < \frac{1}{\sum_{i=0}^{e} \binom{n}{i}}$$

with $r = 2e + 1$.

Proof. For a given $x \in \{0,1\}^n$, let us define the ball

$$B(x, e) = \{y \in \{0,1\}^n \ : \ d_1(x, y) \le e\}.$$

The number of elements in $B(x, e)$ is

$$|B(x, e)| = \sum_{i=0}^{e} \binom{n}{i}.$$

If we have a code $\mathcal{E} = \{x_1, \ldots, x_m\}$ of minimum distance $r = 2e + 1$, then the balls $B(x_i, e)$ are disjoints and thus we have the packing inequality known as *Hamming bound*:

$$m|B(x, e)| \le 2^n.$$

On the other hand, if $\mathcal{E}$ is non-extensible then for every $x \in \{0,1\}^n$ there exists $y \in \mathcal{E}$ such that $d(x, y) \le 2e$. Thus from Theorem 9.1 we get

$$m|B(x, 2e)| \ge 2^n$$

and the corresponding lower bound holds. $\square$

Due to the fundamental importance of Hamming distance for information science many studies have been done towards improving the Hamming bound. A code that attains Hamming bound is called a perfect code, the list of perfect code has been determined [Heden (1975); Van Lint (1975)]:

(1) $(n, 2^n, 1)$ codes for $n \ge 1$.
(2) $(2k + 1, 2, 2k + 1)$ repetition codes for $k \ge 0$.

(3) $(2^k - 1, 2^{2^k - k - 1}, 3)$ codes for $k \geq 2$.
(4) the $(23, 4096, 7)$ Golay code.

More sophisticated upper bounds than the Hamming bound can be obtained. In [Mounits, Etzion and Litsyn (2002)] some upper bounds obtained by a relatively elementary method are given. A more sophisticated technique for getting upper bounds is the linear programming method, which relies on Harmonic Analysis [Conway and Sloane (1999); Mounits, Etzion and Litsyn (2007)]. Yet another technique is the semidefinite programming bounds that relies on yet more algebra and requires more computational effort [Schrijver (2005)]. Another way to strengthen the Hamming bound is based on Information Theoretic methods [Wyner (1964); Ash (1965)]: If $r/n < \frac{1}{2}$, then we have

$$ \delta_{n,r} \leq \frac{n \cdot K(p)}{\sum_{l=0}^{[tpn]} \binom{n}{l} \left(\frac{tr}{2} - l \right)} $$

where $p = r/2n$, $t = (1/2p)(1 - \sqrt{1 - 4p})$ and $K(p) = p/\sqrt{1 - 4p}$.

Consider a set of 2^n points whose coordinates are 1 or 0 in a Euclidean space of dimension n. Euclidean distance is defined between two points of the 2^n points. Our model may be called random sequential coding. Consider a random sequential packing into the 2^n points. At first we choose one point (d coordinates) at random and we record it. Choose another and record it if its Hamming distance is $\geq r$, ($r < n$), otherwise discard it. Now, choose the next point at random and record it if the Hamming distance from each of the two points is not less than r, otherwise discard it and choose another point at random. We continue this procedure until there is no possible point to record among the 2^n points and we now have the number of recorded points $X(n, r)$, which is a random variable.

Define the packing density by $X(n, r)/2^n$ and define $\Delta_{n,r} = E(X(n, r)/2^n)$. Computer simulations suggest that $\Delta_{n+1,r}/\Delta_{n,r}$ approaches 1 as n tends to ∞, as in the case of the simplest cubic random packing. Note that by Theorem 9.2, if the limit of $\Delta_{n+1,r}/\Delta_{n,r}$ exists then it should be 1.

The result of simulations are shown in Table 9.3 for $r = 2$ and 3. Further computational results in [Itoh and Solomon (1986); Itoh (1985)] show that the variance of $X(n, r)$ is large for r even and small for r odd. The form $n^{-\alpha}$ fits our random packing density reasonably well for $r = 2, 3$ where

$\alpha > 0$ is an empirical constant estimated by the least-squares method, see Table 9.3. For the case $r \geq 4$, $n^{-\alpha}$ does not fit our simulation results.

9.5 Frequency of getting Golay code by a random sequential packing

A code is said to be *linear* if its set of codewords is the set of elements of a vector space over the field with two elements $GF(2)$. Linear codes are of great interest since the linear structure allows for many results to be proved for them and faster algorithms. The Golay code $\mathcal{C}_{24}$ is a remarkable $(24, 4096, 8)$ code in dimension 24 with minimum distance 8. It is linear, that is it is a vector space of dimension 12, which thus has

$$|GF(2)|^{12} = 2^{12} = 4096$$

codewords.

This code was discovered in [Golay (1949)] and is applied in a large variety of scientific applications of coding theory. One should point out that even if this code has very good performance, one drawback is that the decoding is not necessarily easy. Around the codeword 0^{24} the number of points at distance 0, 8, 12, 16, resp. 24 is 1, 759, 2576, 759, resp. 1 that is the length distribution is $0^1, 8^{759}, 12^{2576}, 16^{759}, 24^1$. One can prove that this is the best code in dimension 24 with minimum length 8, that is any code of dimension 24 with minimum distance 8 is equivalent to the Golay code by an automorphism of the hypercube $\{0, 1\}^{24}$ (see for example [Delsarte and Goethals (1975)]).

For any $v \in \mathcal{C}_{24}$ the translation operation $t_v(x) = x + v$ preserves $\mathcal{C}_{24}$ since it is a vector space. But in addition there are permutations of the 24 coordinates that also preserve $\mathcal{C}_{24}$. Those permutations define a finite simple group named the Mathieu group M_{24} which is a 5-transitive permutation group (see [Conway and Sloane (1999)] for more details). The automorphism group of $\mathcal{C}_{24}$ has thus order $2^{12}|M_{24}|$. Note that the $(23, 4096, 7)$ Golay code is obtained by dropping one coordinate from the Golay code $\mathcal{C}_{24}$.

On the vector space $\{0, 1\}^n$ it is possible to define a scalar product by

$$\langle x, y \rangle = \sum_{i=1}^{n} x_i y_i.$$

Using this notion we can define the dual of a linear code C to be

$$C^{\perp} = \{x \in \{0, 1\}^n : \langle x, y \rangle \equiv 0 \pmod{2} \text{ for all } y \in C\}.$$

The dual code $C^\perp$ is linear of dimension $n - \dim(C)$. The dual of the Golay code $\mathcal{C}_{24}$ is another code that is isomorphic to $\mathcal{C}_{24}$, thus $\mathcal{C}_{24}$ is called a *self-dual* code. The doubly even (that is ones for which $\langle x, y \rangle \in 4\mathbb{Z}$ for $x, y \in C$) self-dual codes of $\{0,1\}^{24}$ have been enumerated in [Conway and Pless (1980)]. Their possible weight distribution are of the form

0	4	8	12	16	20	24
1	$6k$	$759 - 24k$	$2576 + 36k$	$759 - 24k$	$6k$	1

with $k = 0$, 1, 2, 3, 4, 5, 7, 11. We should point out that the Golay code allows one to build the Leech lattice which is the packing of highest density in dimension 24 and has many remarkable properties [Conway and Sloane (1999); Leech (1964)]. See [MacWilliams and Sloane (1977); Thompson (1983); Conway and Sloane (1999); Peterson (1961); Bannai and Sloane (1981); Mallows, Pless and Sloane (1976)] for proofs and more results on this remarkable 24-dimensional code and on its siblings.

All that being said, it is not so easy to define $\mathcal{C}_{24}$ explicitly. Several constructions exist, which is to be expected since the code has so many remarkable properties but most of them depend on some algebra. Here we give a sequential random packing model to generate the Golay code $\mathcal{C}_{24}$. If one applies the classical sequential random packing procedure then when adding a vector, we consider only the vectors which have distance in $\{8, \ldots, 24\}$ with all other vectors. A useful restriction in the random packing procedure is to consider random packing and add new codewords only if they have distance 8, 12, 16 or 24 with all other codewords. By doing this random packing with this restriction, 11 trials out of 550 trials produced a code with 4096 elements, that is the Golay code [Itoh (1986); Itoh and Jimbo (1987)].

This is quite remarkable, but not an isolated fact when working with remarkable geometrical structures: for example in [Ballinger et al. (2009)] it has been found that one obtains with probability 90% the kissing configuration of the E_8 lattice when searching for kissing configurations in dimension 8 with 240 vectors. This indicates a direction of research which is starting to develop: look after remarkable geometrical structures by using techniques from optimization and simulated annealing, see [Ballinger et al. (2009)] or some of the later chapters for some illustration of this.

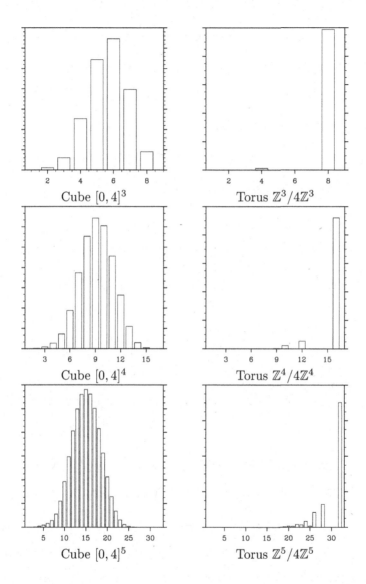

Fig. 9.3 Repartition of number of cubes obtained by sequential random cube packing in the cube $[0,4]^n$ and the torus $\mathbb{Z}^n/4\mathbb{Z}^n$ for $n = 3$, 4 and 5

Chapter 10

Discrete cube packings in the cube

We consider the simplest random sequential packing of higher dimension. The 2-cubes, integral translates of the cube $K_2 = [-1, 1]^n$, are placed (parked) into the cube $K_4 = [-2, 2]^n$ sequentially at random. All the centers of 2-cubes are integer points in K_4 and all admissible positions at each step are equiprobable. The process continues until there is no space to place (park), namely until saturation. The problem is to get the expectation M of number of 2-cubes in the cube $K_4 = [-2, 2]^n$ at the saturation. The computer simulations up to dimension 11, suggest the packing density is around $n^{-\alpha}$, i.e. that the number of cubes is on average around

$$\frac{2^n}{n^\alpha}$$

with an appropriate empirical constant α, as in the case of random sequential coding by Hamming distance [Itoh and Ueda (1983); Itoh and Solomon (1986)]. The expectation of number of cubes at the saturation is mathematically shown to be not less than $\left(\frac{3}{2}\right)^n$ [Poyarkov (2005)], which is the first non-trivial exponential lower bound for the random sequential packing of the dimension higher than one. This Chapter is adapted from [Poyarkov (2005)] and also from [Dolbilin, Itoh and Poyarkov (2005)].

10.1 Setting of a goal

Consider the n-dimensional cube $K_4 = \{y \in \mathbb{R}^n : -2 \le y_i \le 2\}$. The length of its edges is 4. We shall take cubes with the edge length 2 parallel to the edge of the K_4 and call them 2-cubes. Therefore the centers y^1 and y^2 of two 2-cubes of a packing satisfy the following relations:

(1) $| y_i^1 | \leq 1$ and $| y_i^2 | \leq 1$ for $1 \leq i \leq n$,

(2) $\max_{1 \leq i \leq n} | y_i^1 - y_i^2 | = 2$.

We consider only integral packings, i.e., such that the coordinates of 2-cubes are equal to 0 or ± 1. An integral packing into K_4 is said to be saturated if we cannot add a new 2-cube without overlapping with the previous ones.

We give two examples of packings.

Take all the 2-cubes such that all their centers have coordinates equal to ± 1. These 2^n cubes obviously do not overlap and form an integer-valued packing of the K_4 cube. Since these cubes cover the cube K_4 totally, the integral packing is saturated.

Take a packing P which consists of only the 2-cube C with center $\{0, 0, \ldots, 0\}$. Since any 2-cube located in the K_4 and C overlap, the packing P is also saturated.

Problem 10.1. *(Random sequential integral packing into the K_4-cube by 2-cubes [Itoh and Ueda (1983)]). 2-cubes are placed (parked) sequentially at random into the cube K_4 until there is no space to place without overlapping, that is to say, until saturation. Here all their centers are integer points in K_4 and all admissible positions at each step are equiprobably chosen by the random number independently distributed from the previous steps. Find the expectation M of number of 2-cubes in the cube K_4 at the saturation.*

We have just seen that the number of cubes in a random saturated packing varies from 1 to 2^n. We prove the following theorem.

Theorem 10.1. *[Poyarkov (2005)] The expectation M of number of 2-cubes obtained by random sequential integral packing into the 4-cube K_4 by 2-cubes at the saturation is not less than $(3/2)^n$.*

10.2   Reduction to another problem

Let us formulate another, equivalent problem to prove the theorem. The cube $K_4 = [-2, 2]^n$ has faces of dimension 0 to n. Every face of dimension k is itself a k-dimensional cube and thus contains 2^k vertices. The face of dimension n is the cube K_4 itself. The number of faces of dimension k is $2^{n-k} \binom{n}{k}$. So, for $k = 0, 1, n-1, n$ it is 2^n, $2^{n-1}n$, $2n$ and 1. The group acting on the cube is the Coxeter group B_n of order $2^n n!$, that is the semidirect product of the group $\mathbb{Z}_2^n$ with the symmetric group $\text{Sym}(n)$.

Problem 10.2. *(Random sequential face removing [Poyarkov (2005)]).* *For a given d-dimensional cube $K_4 = \{y \in \mathbb{R}^n : -2 \le y_i \le 2\}$, at the first step choose a closed face (of arbitrary dimension) with equal probability $(1/3)^n$ and remove the face. At each step we choose with equal probability a face among remained (after the previous removals) faces of any dimension i for $0 \le i \le n$ and remove it. Here i-dimensional face F^i means the closure of the face, that includes all faces incident to the face F^i of lower dimension up to vertices of F^i. The steps continue until the last closed face will disappear. Find the expectation M' of number of removed faces at the termination.*

A link between Problem 10.1 and Problem 10.2 is seen from the following Lemma 10.1.

Lemma 10.1. *We have $M = M'$.*

Proof. The center $c^i = (c^i_j)_{1 \le j \le n}$ of each 2-cube in the K_4 has coordinates -1, 0, or 1. Each face F^k of K_4 with dimension k, $0 \le k \le n$ has a center $c \in \{-1, 0, 1\}^n$ and conversely each point $p \in \{-1, 0, 1\}^n$ is the center of the only face of the cube K_4 of some dimension k which is the number of 0-coordinates.

Thus, there is a bijection between faces of the cube K_4 and positions of 2-cubes in the cube K_4.

Hence there is a bijection between a sequential face removing and the corresponding sequential packing.

Two faces F^k and F^l have no common vertices (do not intersect each other at all) if and only if the corresponding 2-cubes do not overlap. Two 2-cubes do overlap if and only if for the centers c and c', $c_i c'_i \ne -1$ for all i.

Hence the probability of getting a sequence of cubes by random packing is equal to the probability of getting a corresponding sequence of faces obtained by the bijection. Hence the expectation of number of small 2-cubes in a random sequential packing into the cube K_4 at the saturation is equal to the expectation of number of removed faces in the cube K_4 at the termination. □

The correspondence between Problem 10.1 and Problem 10.2 is illustrated in Figure 10.1. Now we obtain the expectation of number of faces removed at the first step.

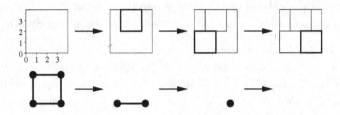

Fig. 10.1　The 2-dimensional case of the simplest sequential random cube packing process and the corresponding process on the face lattice of the cube

Lemma 10.2. *The expectation V_1 of number of vertices of the cube K removed at the first step is $\left(\frac{4}{3}\right)^n$.*

Proof. The total number of faces in a cube is 3^n. The number of faces with dimension k is $2^{n-k}\binom{n}{k}$. In removing a k-dimensional face, 2^k vertices are removed. Now it is easy to count the mean number V_1:

$$V_1 = \frac{1}{3^n}\sum_{k=0}^{n} 2^{n-k}\binom{n}{k}2^k = \frac{2^n}{3^n}\sum_{k=0}^{n}\binom{n}{k} = \left(\frac{4}{3}\right)^n.$$

The lemma is proved. □

Now we remove faces from the cube K_4. At each step of the process, we obtain some remains of the cube. Let us denote by N_k the number of complete (i.e. those that may be removed) faces of dimension k.

Lemma 10.3. *Let $k \leq l$ then*

$$N_l \leq \frac{\binom{n}{l}}{2^{l-k}\binom{n}{k}}N_k.$$

Proof. Let us denote the number of k-faces lying in N_l l-faces by N. Then $N = 2^{l-k}\binom{l}{k}N_l$. But every k-face lies in not more than S l-faces, where S is the number of l-faces in a complete cube that have a fixed k-face. Then

$$S = \frac{2^{n-l}\binom{n}{l}\,2^{l-k}\binom{l}{k}}{2^{n-k}\binom{n}{k}} = \frac{\binom{n}{l}\,\binom{l}{k}}{\binom{n}{k}}.$$

Hence we have

$$N_k \geq \frac{N}{S} = \frac{2^{l-k}\binom{l}{k}N_l}{\frac{\binom{n}{l}\binom{l}{k}}{\binom{n}{k}}} = \frac{2^{l-k}\binom{n}{k}}{\binom{n}{l}}N_l.$$

The lemma is proved. □

Lemma 10.4. *Let $(b_i)_{i=0}^n$ be a sequence of non-negative numbers and $B = \sum_{i=0}^n b_i$. For fixed k, $0 \le k \le n$, let the sequence $(b_i)_{i=0}^n$ satisfy the following two conditions:*

(1) for any $i < k$, we have

$$b_i = \left(\frac{2}{3}\right)^n \frac{\binom{n}{i}}{2^i} B,$$

(2) for any i and j with $k \le i < j \le n$, the following inequality holds:

$$b_j \le \frac{\binom{n}{j}}{2^{j-i}\binom{n}{i}} b_i.$$

Then

$$b_k \ge \left(\frac{2}{3}\right)^n \frac{\binom{n}{k}}{2^k} B.$$

Proof. We can write

$$\left(\tfrac{2}{3}\right)^n B \left(\left(\tfrac{3}{2}\right)^n - \sum_{i=0}^{k-1} \frac{\binom{n}{i}}{2^i}\right) = B - \sum_{i=0}^{k-1} b_i$$

$$= \sum_{i=k}^n b_i$$

$$\le \sum_{i=k}^n \frac{\binom{n}{i}}{2^{i-k}\binom{n}{k}} b_k$$

$$= \frac{2^k b_k}{\binom{n}{k}} \sum_{i=k}^n \frac{\binom{n}{i}}{2^i}.$$

But since

$$\sum_{i=0}^n \frac{\binom{n}{i}}{2^i} = \left(\frac{3}{2}\right)^n$$

we obtain

$$\left(\frac{3}{2}\right)^n - \sum_{i=0}^{k-1} \frac{\binom{n}{i}}{2^i} = \sum_{i=k}^n \frac{\binom{n}{i}}{2^i}.$$

So, we have

$$b_k \ge \left(\frac{2}{3}\right)^n \frac{\binom{n}{k}}{2^k} B.$$

We are done. $\square$

Lemma 10.5. *The expectation V of number of removed vertices of the cube K at each step is not larger than $V_1 = \left(\frac{4}{3}\right)^n$.*

Proof. Note that N_k is the number of complete faces of dimension k remaining before the next face removal. Hence, the expectation V of number of removed vertices is equal to

$$V = \frac{1}{N} \sum_{i=0}^{n} 2^i N_i.$$

Now let

$$\varphi(x) = \frac{1}{N} \sum_{i=0}^{n} 2^i x_i.$$

Let us note that if with this k, $1 \le k \le n$, the following conditions hold:

(1) $\sum_{i=0}^{n} x_i^1 = \sum_{i=0}^{n} x_i^2$,
(2) $x_i^1 \ge x_i^2$ if $0 \le i < k$,
(3) $x_i^1 \le x_i^2$ if $k \le i \le n$,

then obviously $\varphi(x^1) \le \varphi(x^2)$.

Let $x_i^0 = N_i$, $0 \le i \le n$. Then the expectation of number of removed vertices at each step is equal to $\varphi(x^0)$. Let us note that from Lemma 10.3, to numbers x_i^0 the following condition applies. For $i < j \le n$, we have

$$x_j^0 \le \frac{\binom{n}{j}}{2^{j-i}\binom{n}{i}} x_i^0.$$

Applying Lemma 10.4 to x^0 and $k = 0$, we see that

$$x_0^0 \ge \left(\frac{2}{3}\right)^n \frac{\binom{n}{0}}{2^0} N = \left(\frac{2}{3}\right)^n N.$$

Let

$$x_0^1 = \left(\frac{2}{3}\right)^n \frac{\binom{n}{0}}{2^0} N, \quad x_1^1 = x_1^0 + x_0^0 - x_0^1, \text{ and } x_i^1 = x_i^0 \text{ for } i \ne 0, 1.$$

Then

$$\sum_{i=0}^{n} x_i^1 = \sum_{i=0}^{n} x_i^0 = N.$$

Hence $\varphi(x^1) \ge \varphi(x^0)$. Moreover, for any i, j with $0 \le i < j \le n$, we have

$$x_j^1 \le \frac{\binom{n}{j}}{2^{j-i}\binom{n}{i}} x_i^1.$$

Applying Lemma 10.4 for the set x^1 and $k = 1$, we obtain that

$$x_1^1 \ge \left(\frac{2}{3}\right)^n \frac{\binom{n}{1}}{2^1} N.$$

Let

$$x_1^2 = \left(\frac{2}{3}\right)^n \frac{\binom{n}{1}}{2^1} N, \quad x_2^2 = x_2^1 + x_1^1 - x_1^2, \text{ and } x_i^2 = x_i^1 \text{ for } i \neq 1, 2.$$

Then

$$\sum_{i=0}^{n} x_i^2 = \sum_{i=0}^{n} x_i^1 = N.$$

Hence $\varphi(x^2) \geq \varphi(x^1)$. In addition, for any i, j with $1 < i < j \leq n$, we have

$$x_j^2 \leq \frac{\binom{n}{j}}{2^{j-i}\binom{n}{i}} x_i^2.$$

Applying Lemma 10.4, one iterates and finally obtains

$$x_i^n = \left(\frac{2}{3}\right)^n \frac{\binom{n}{i}}{2^i} N,$$

and

$$V = \varphi(x^0) \leq \varphi(x^1) \leq \cdots \leq \varphi(x^n) = \left(\frac{4}{3}\right)^n.$$

We are done. □

10.3 Proof of the theorem

In Lemma 10.5, we have proved that, at each moment, the expectation of number of vertices removed from the cube K is not larger than $V_1 = (4/3)^n$. There are only 2^n vertices in the cube K, and at the end of the process all of them are removed. Hence intuitively it is natural to conclude that

$$M = M' \geq \frac{2^n}{(4/3)^n} = \left(\frac{3}{2}\right)^n. \tag{10.1}$$

We prove inequality (10.1) by using the following Lemmas 10.6-10.9.

Consider all possible states of the cube K during the process of removing vertices (every state is described unambiguously by the subset of non-removed vertices). Since we remove only closed faces along with all vertices spanning the faces, we get that any state is uniquely determined by some subset of vertices of the cube. A subset of vertices of the cube determines a state of the cube uniquely. It is obvious that the number of all possible states of the cube is $L \doteq 2^{2^n}$. Let us denote these states by $A_1, A_2, \ldots, A_L$. The initial state A_1 corresponds to the full cube with all the 2^n vertices. All vertices are removed from the cube in the A_L. Let α_i

be the probability of the appearance of state A_i (the probability of passing vertex A_i) in the process of removing faces from the cube K.

Consider the graph G whose vertices are states of the cube K. If it is possible to obtain state A_j by removing one of the residuary closed faces, we connect two vertices A_i and A_j in the graph G by a directed edge e_{ij}. Let us denote the probability of the traversal through edge e_{ij} (i.e. the probability of getting state A_i after removing faces from the cube K and having state A_j after removing the next face) by β_{ij}. We denote the number of vertices and the set of vertices in the face removed by this transition by K_{ij} and D_{ij} respectively. K_{ij} is equal to the disparity between the number of vertices in states A_i and A_j.

Any concrete sequence of removals of faces can be interpreted as an edge path in the graph of states passing through vertices and directed edges of the graph and connecting the initial state to the final one.

Let G_l be the set of all possible directed graphs of g, which have a set of $T_l = l$ vertices $\{A(1), A(2), \ldots, A(l)\}$ of the graph G, and a set of $T_l - 1$ directed edges $\{e_{A(1)A(2)}, \ldots, e_{A(i)A(i+1)}, \ldots, e_{A(l-1)A(l)}\}$, which make a sequence of one-by-one face removing, from the initial state A_1 corresponding to the full cube to the final state A_L which has no vertex. Note that $A(1) = A_1$ and $A(l) = A_L$. Let G^* be the set of all possible sets G_l. Let the directed graph g be with the probability Π_g.

We see that

$$\sum_{G_l \in G^*} \sum_{g \in G_l} \Pi_g = 1$$

and

$$\sum_{G_l \in G^*} \sum_{g \in G_l} T_l \Pi_g = M.$$

Define an indicator for two graphs X and Y as,

$$I_X(Y) = \begin{cases} 1 \text{ if } Y \subseteq X, \\ 0 \text{ otherwise.} \end{cases}$$

By simple summation, we get

$$\alpha_i = \sum_{G_l \in G^*} \sum_{g \in G_l} I_g(A_i) \Pi_g$$

and

$$\beta_{ij} = \sum_{G_l \in G^*} \sum_{g \in G_l} I_g(e_{ij}) \Pi_g.$$

Lemma 10.6. *We have*

$$\alpha_i = \sum_{A_j \subset G} \beta_{ij}.$$

Proof. We have the equality

$$\sum_{A_j \subset G} \beta_{ij} = \sum_{G_l \in G^*} \sum_{g \in G_l} \sum_{A_j \subset G} I_g(e_{ij}) \Pi_g$$
$$= \sum_{G_l \in G^*} \sum_{g \in G_l} I_g(A_i) \Pi_g$$
$$= \alpha_i.$$

The result follows. □

Lemma 10.7. *We have*

$$\sum_{e_{ij} \subset G} K_{ij} \beta_{ij} \leq \left(\frac{4}{3}\right)^n \sum_{e_{ij} \subset G} \beta_{ij}.$$

Proof. Since the probability of getting state A_j from state A_k with the condition of being in state A_k is β_{kj}/α_k, by applying Lemma 10.6, we have from Lemma 10.5,

$$\frac{\sum_{A_j \subset G} K_{ij} \beta_{ij}}{\sum_{A_j \subset G} \beta_{ij}} \leq \left(\frac{4}{3}\right)^n.$$

The result follows. □

Lemma 10.8. *We have*

$$\sum_{e_{ij} \subset G} \beta_{ij} K_{ij} = 2^n.$$

Proof. Since any $A_k \subset G$ should be removed, the sum of probability is

$$\sum_{e_{ij} \subset G} \sum_{G_l \in G^*} \sum_{g \in G_l} I_{D_{ij}}(A_k) I_g(e_{ij}) \Pi_g = 1.$$

Hence we have

$$\sum_{A_k \subset G} \sum_{e_{ij} \subset G} \sum_{G_l \in G^*} \sum_{g \in G_l} I_{D_{ij}}(A_k) I_g(e_{ij}) \Pi_g = 2^n.$$

Since

$$\sum_{A_k \subset G} \sum_{e_{ij} \subset G} \sum_{G_l \in G^*} \sum_{g \in G_l} I_{D_{ij}}(A_k) I_g(e_{ij}) \Pi_g$$
$$= \sum_{e_{ij} \subset G} \sum_{G_l \in G^*} \sum_{g \in G_l} \sum_{A_k \subset G} I_{D_{ij}}(A_k) I_g(e_{ij}) \Pi_g$$
$$= \sum_{e_{ij} \subset G} \sum_{G_l \in G^*} \sum_{g \in G_l} K_{ij} I_g(e_{ij}) \Pi_g$$
$$= \sum_{e_{ij} \subset G} K_{ij} \sum_{G_l \in G^*} \sum_{g \in G_l} I_g(e_{ij}) \Pi_g$$
$$= \sum_{e_{ij} \subset G} K_{ij} \beta_{ij},$$

the result follows.   □

Lemma 10.9. *We have*

$$\sum_{A_i \subset G} \alpha_i = M,$$

where M is the expectation of number of faces which are removed unless the final state A_L is attained.

Proof.   Since for $g \in G_l$

$$\sum_{A_i \subset G} I_g(A_i) = T_l,$$

and

$$\alpha_i = \sum_{G_l \in G^*} \sum_{g \in G_l} I_g(A_i) \Pi_g,$$

we have

$$\sum_{A_i \subset G} \alpha_i = \sum_{A_i \subset G} \sum_{G_l \in G^*} \sum_{g \in G_l} I_g(A_i) \Pi_g$$

$$= \sum_{G_l \in G^*} \sum_{g \in G_l} \sum_{A_i \subset G} I_g(A_i) \Pi_g$$

$$= \sum_{G_l \in G^*} \sum_{g \in G_l} T_l \Pi_g = M$$

which is what needed to be proved.   □

Now Theorem 10.1 follows immediately by using the above lemmas.

$$M = \sum_{A_i \subset G} \alpha_i = \sum_{e_{ij} \subset G} \beta_{ij} \geq \left(\frac{3}{4}\right)^n \sum_{e_{ij} \subset G} \beta_{ij} K_{ij} = \left(\frac{3}{4}\right)^n 2^n = \left(\frac{3}{2}\right)^n.$$

Chapter 11

Discrete cube packings in the torus

We consider tilings and packings of $\mathbb{R}^n$ by integral translates of cubes $[0,2)^n$, which are $4\mathbb{Z}^n$-periodic. Such cube packing are the direct analog of the simplest random sequential cube packing model for the cube $[0,4]^n$. Such cube packings were already considered in the study of the Keller conjecture which asserts that in any cube tiling there is at least one face-to-face adjacency (this was proved to be false). Such cube packings can be described by cliques of an associated graph, which allow us to classify them in dimensions $n \leq 4$. For higher dimensions, we use random methods to generate some examples.

Such a cube packing is called *non-extensible* if we cannot insert a cube in the complement of the packing. In dimension 3, there is a unique non-extensible cube packing with 4 cubes. We prove that n-dimensional cube packings with more than $2^n - 3$ cubes can be extended to cube tilings. We also give a lower bound on the number N of cubes of non-extensible cube packings.

Given such a cube packing and $z \in \mathbb{Z}^n$, we denote by N_z the number of cubes inside the 4-cube $z + [0,4)^n$ and call *second moment* the average of N_z^2. We prove that the regular tiling by cubes has maximal second moment and give a lower bound on the second moment of a cube packing in terms of its density and dimension.

11.1 Introduction

A *general cube tiling* is a tiling of $\mathbb{R}^n$ by translates of the hypercube $[0,2)^n$, which we call a 2-cube. A *special cube tiling* is a tiling of $\mathbb{R}^n$ by integral translates of the hypercube $[0,2)^n$, which are $4\mathbb{Z}^n$-periodic. An example of such a tiling is the *regular cube tiling* of $\mathbb{R}^n$ by cubes of the form $z + [0,2)^n$

with $z \in 2\mathbb{Z}^n$.

In dimension 1, there is only one type of special cube tiling, while in dimension 2, two types of special cube tilings exist (see section 11.2 for the classification methodology): The Keller's cube tiling conjecture (see [Keller (1930)]) asserts that any tiling of $\mathbb{R}^n$ by translates of a unit cube admits at least one face-to-face adjacency. It is proved in [Szabó (1986)] that if this conjecture has a counterexample, then there is another counterexample, which is also a special cube tiling. Using this, the Keller conjecture was solved negatively for $n \geq 10$ in [Lagarias and Shor (1992)] and $n \geq 8$ in [Mackey (2002)] (note that the conjecture is proved to be true for $n \leq 6$ in [Perron (1940)]). Hence, special cube tilings, while seemingly limited objects have a lot of combinatorial possibilities. In the rest of this chapter *cube tiling* stands for special cube tilings and N is the number of orbits of cubes under the translation group $4\mathbb{Z}^n$. Another equivalent viewpoint is to say that we are doing tilings of the torus $\mathbb{R}^n/4\mathbb{Z}^n$ and N is then the number of cubes in this torus.

A *cube packing* is a $4\mathbb{Z}^n$-periodic set of integral translates of the 2-cube, such that any two cubes are non-intersecting. A cube packing is called *non-extensible* if one cannot insert any more cubes. Starting from dimension 3, there are non-extensible cube packings, which are not cube tilings (this first appear in [Lagarias, Reeds and Wang (2000)]). In dimension 3 this cube packing is unique (see Figure 11.1) and it is the source of much of the inspiration of this chapter.

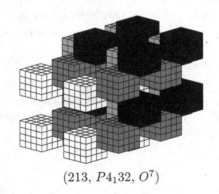

$$(213, P4_132, O^7)$$

Fig. 11.1   The unique non-tiling non-extensible cube packing in dimension 3

In Section 11.2, following [Lagarias and Shor (1992)], we present a translation of the packing and tiling problems into clique problems in graphs.

Explicit methods, in GAP, are used up to $n = 4$. For $n \geq 5$, we use various random methods, in Fortran 90 for generating random cube packings.

Denote by $f(n)$ the smallest number of cubes, which form a non-extensible cube packing. In Section 11.3, we give some lower and upper bounds on the value of $f(n)$. In [Dolbilin, Itoh and Poyarkov (2005)] it is proved that any cube packing of $[0, 4)^n$ by cubes $[0, 2)^n$ is extensible to a $4\mathbb{Z}^n$-periodic cube tiling of $\mathbb{R}^n$.

If $\mathcal{CP}$ is a cube packing, denote by $hole(\mathcal{CP})$ and call *hole*, its complement $\mathbb{R}^n - \mathcal{CP}$. We prove that if a cube packing has more than $2^n - 3$ cubes, then it is extensible to a tiling, i.e. that holes of volume at most 3 are fillable. We also obtain some conjectures on non-fillable holes of volume at most 7.

Given a cube packing $\mathcal{CP}$, the *counting function* $N_z(\mathcal{CP})$ is defined as the number of cubes of $\mathcal{CP}$ contained in $z + [0, 4)^n$. We study its second moment in Section 11.4. We prove that the highest second moment for tilings is attained for the regular cube tiling and give a lower bound for the second moment of cube packings, in terms of its dimension n and number of cubes N.

11.2 Algorithm for generating cube packings

Every 2-cube of an n-dimensional cube packing is equivalent under $4\mathbb{Z}^n$ to a cube with center in $\{0, 1, 2, 3\}^n$. Two 2-cubes of centers x and x' do not overlap if and only if there exists a coordinate i, such that $|x_i - x_i'| = 2$. So, one considers the graphs G_n (introduced in [Corrádi and Szabó (1990)]) with vertex-set $\{0, 1, 2, 3\}^n$ and two vertices being adjacent if and only if their associated cubes do not overlap. Cube packings correspond to cliques of G_n; they are non-extensible if and only if the cliques are maximal. Cube tilings correspond to maximum cliques (i.e. cliques of maximum size 2^n). Actually such cube packing graphs and Keller cube packing graphs are used for benchmarking clique enumeration programs.

For a given n, the graph G_n has a finite number of vertices and an automorphism group $\text{Aut}(G_n)$ of size $8^n n!$. Hence, it is theoretically possible to do the enumeration of the cliques of G_n. The algorithm consists of using the set of all cliques with N vertices, considering all possibilities of extension, and then reducing by isomorphism using $\text{Aut}(G_n)$ (the actual computation was done in GAP, see [Gap (2002)]). The group $\text{Aut}(G_n)$ is presented as a permutation group in GAP and the cliques as subsets of $\{1, \ldots, 4^n\}$. GAP uses backtrack search for testing if two subsets are equivalent under $\text{Aut}(G_n)$ and is hence, very efficient even for large values of n. This enumeration is, in practice, possible only for $n \leq 4$ due to the huge number of cliques that appear.

For $n = 2$, one finds only two orbits of maximal cliques of 4 vertices, i.e. two cube tilings. See Figure 11.2 for their picture. For $n = 3$, there is a unique orbit of maximal clique with 4 vertices, while there are 9 orbits of maximum cliques (i.e. cube tilings). See Figure 11.3 for their picture. We follow the classical methodology for the space group of those tilings, see Appendix 11.5 for its description.

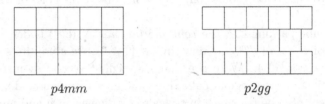

$p4mm$ $p2gg$

Fig. 11.2   The 2 isomorphism type of cube tilings in dimension 2

For $n = 4$, the computations are still possible and one finds the following results with N being the number of vertices of the maximal clique.

N	1	2	3	4	5	6	7	8	9	10	11	12	13	14	15	16
# orbit maximal cliques	0	0	0	0	0	0	0	38	6	24	0	71	0	0	0	744

Suppose that we have a cube tiling with two cube centers x and x', satisfying $x' = x + 2e_i$ with e_i being the i-th unit vector, i.e. they have a face-to-face adjacency. If one replace x, x' by $x + e_i$, $x' + e_i$ and leave other centers unchanged, then one obtains another cube tiling, which we call the *flip* of the original cube tiling (see an example on Figure 11.4).

The enumeration strategy is then the following: take as initial list of orbits the orbit of the regular cube tiling. For every orbit of cube packing, compute all possible pairs $\{x, x'\}$, which allow to create a new cube tiling. If the corresponding orbits of cube tilings are new, then we insert them into the list of orbits. Given a dimension n, consider the graph Co_n, whose vertex-set consists of all orbits of cube tilings and put an edge between two orbits if one is obtained from the other by a flipping. The above algorithm consists of computing the connected component of the regular cube tiling in Co_n. Since the Keller conjecture is false in dimension $n \geq 8$, we know that in those dimensions there are some isolated vertices in the

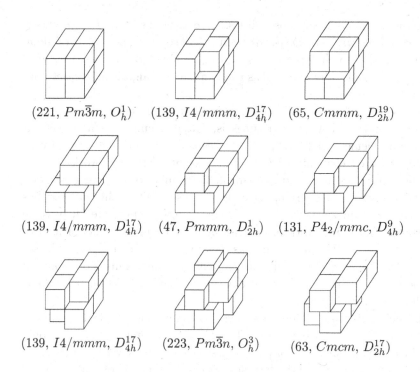

$(221, Pm\bar{3}m, O_h^1)$ $(139, I4/mmm, D_{4h}^{17})$ $(65, Cmmm, D_{2h}^{19})$

$(139, I4/mmm, D_{4h}^{17})$ $(47, Pmmm, D_{2h}^1)$ $(131, P4_2/mmc, D_{4h}^9)$

$(139, I4/mmm, D_{4h}^{17})$ $(223, Pm\bar{3}n, O_h^3)$ $(63, Cmcm, D_{2h}^{17})$

Fig. 11.3   The 9 isomorphism type of cube tilings in dimension 3

Fig. 11.4   Example of flip for a 3 dimensional cube tiling

graph and so, the above algorithm does not work. However, the graph Co_n is connected for $n \le 4$, i.e. any two cube tilings in those dimensions can be obtained by a sequence of flipping. It is an interesting question to decide, in which dimension n the graph Co_n is connected; the only remaining unsolved dimensions are $n = 5, 6, 7$.

For dimensions $n \ge 5$, two above enumeration methods cannot work since there are too many possibilities. Hence, we used random methods. The *sequential random cube packing* consists of selecting points, at random, on $\{0, 1, 2, 3\}^n$, so that the corresponding 2-cubes do not intersect, until one cannot do this any more. This random packing algorithm creates non-extensible cube packings.

The actual algorithm for creating non-extensible cube packings is as follows: the list L of selected cubes is, initially, empty. One selects at random an element of $\{0, 1, 2, 3\}^n$ and keeps it if it does not overlap with already selected elements of L. Of course not every trial works and as the space becomes more and more filled, the number of random generation needed to get a non-overlapping cube increase. When this number has reached a certain level, we go to a second stage: enumerate all possible insertable cubes, and work in this list by eliminating elements of it after choices are made. This algorithm has the advantage of enumerating the set $\{0, 1, 2, 3\}^n$ only one time and is hence, relatively fast. This is our solution to the covering problem that we have introduced in Section 9.1. Some results are shown in Figure 9.3.

If one wants to find some packings with low density, then the above strategy is not necessarily the best. The *greedy algorithm* consists of keeping all 4^n elements in memory and at every step generate, say 20 elements and keep the one which cover the largest part of the remaining space.

Another possibility is what we call *Metropolis algorithm* (see [Liu (2001); Diaconis (2008)]), or in other words Monte Carlo Markov Chain: we take a non-extensible cube packing, remove a few cubes and rerun a random generation from the remaining cubes. If obtained packing is better than the preceding one, or not worse than a specified upper bound, then we keep it; otherwise, we rerun the algorithm. This strategy allows one to make a random walk in the space of non-extensible cube packings and is based on the assumption, that the best non-extensible cube packings are not far from other, less good, non-extensible cube packings.

11.3   Non-extensible cube packings

In dimension 1 or 2, any cube packing is extensible to a cube tiling. The exhaustive enumeration methods of the preceding section show that in dimension 3, there is a unique non-extensible cube packing, which is not a tiling. The set of its centers is, up to an automorphism of G_3:

$$\{(0, 0, 0), (3, 2, 3), (2, 1, 1), (1, 3, 2)\}$$

and its corresponding drawing is shown in Figure 11.1. Its space group symmetry is $P4_132$, which is a chiral group.

We first concentrate on the problem of finding non-extensible cube packings with the smallest number $f(n)$ of cubes. From Section 11.2, we know that $f(1) = 2$, $f(2) = 4$, $f(3) = 4$ and $f(4) = 8$.

Lemma 11.1. *[Dutour Sikirić, Itoh and Poyarkov (2007)] For any $n, m \geq 1$, one has the inequality $f(n + m) \leq f(n)f(m)$.*

Proof. Let P_A and P_B be non-extensible cube packings of $\mathbb{R}^n$ and $\mathbb{R}^m$ with $f(n)$ and $f(m)$ cubes, respectively. Let $a^k = (a_1^k, a_2^k, \ldots, a_n^k)$ and $b^l = (b_1^l, b_2^l, \ldots, b_m^l)$ with $1 \leq k \leq f(n)$, $1 \leq l \leq f(m)$ be the centers of the 2-cubes from P_A and P_B.

Define P to be the set of 2-cubes $C^{k,l}$ with centers $c^{k,l} = (a_1^k, a_2^k, \ldots, a_n^k, b_1^l, b_2^l, \ldots, b_m^l)$ for $1 \leq k \leq f(n)$ and $1 \leq l \leq f(m)$. The size of P is $f(n)f(m)$ and it is easy to check that P is a packing.

Take a cube D with center $d = (d_1, d_2, \ldots, d_{n+m})$. The vector $(d_1, \ldots, d_n)$ overlaps with a 2-cube, say A^{k_0} in P_A, while the vector $(d_{n+1}, \ldots, d_{n+m})$ overlaps with a 2-cube, say B^{l_0} in P_B. Clearly, D overlaps with C^{k_0,l_0} and P is non-extensible. $\square$

Since, $f(3) = 4$, one has $f(6) \leq 16$.

A *blocking set* is a set $\{v^j\}$ of vectors in $\{0, 1, 2, 3\}^n$, such that for every other vector v, there exists a j such that the 2-cubes of center v^j and v overlap. Denote by $h(n)$ the minimum size of a blocking set. Clearly, non-extensible cube packings are blocking sets; so, $h(n) \leq f(n)$.

It is easy to see that $h(2) = 3$ and that any blocking sets of size 3 belong to one of the following two orbits:

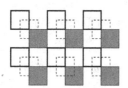 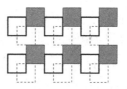

A slightly more complicated computation shows that $h(3) = 4$ and that any blocking set of size 4 belong to one of the following three orbits:

$$\{(0,0,0), (1,1,1), (2,2,2), (3,3,3)\},$$
$$\{(0,0,0), (1,1,1), (2,2,3), (3,3,2)\},$$
$$\{(0,0,0), (3,2,3), (2,1,1), (1,3,2)\}.$$

Lemma 11.2. *[Dutour Sikirić, Itoh and Poyarkov (2007)] Let N satisfy the inequality $\lfloor \frac{3N}{4} \rfloor < h(n)$, then one has $h(n+1) > N$.*

Proof. First $h(n) > N$ if and only if, for any set P of N 2-cubes, there exists a 2-cube D, which does not overlap with any 2-cube from P.

Let P be a set of N 2-cubes in the torus $\mathbb{R}^n/4\mathbb{Z}^n$. Then at least $\lceil \frac{N}{4} \rceil$ centers of them have $x_{n+1} = t$, for some $t \in \{0,1,2,3\}$. Let us define another set P' of vectors by removing those vectors and the $(n+1)$-th coordinate for the remaining vectors. Then P' consists of at most $N - \lceil \frac{N}{4} \rceil = \lfloor \frac{3N}{4} \rfloor$ 2-cubes. But $\lfloor \frac{3N}{4} \rfloor < h(n)$; so, there exists a 2-cube C with center $c = (c_1, c_2, \ldots, c_n)$, which do not overlap with any 2-cube in P'. But then the 2-cube with center $(c_1, c_2, \ldots, c_n, t+2)$ does not overlap with any 2-cube from P. □

Theorem 11.1. *[Dutour Sikirić, Itoh and Poyarkov (2007)] For any $n \geq 1$, one has $h(n+1) \geq \lfloor \frac{4h(n)-1}{3} \rfloor + 1$.*

Proof. Let $N = \lfloor \frac{4h(n)-1}{3} \rfloor$, then it holds:

$$\left\lfloor \frac{3N}{4} \right\rfloor = \left\lfloor \frac{3 \left\lfloor \frac{4h(n)-1}{3} \right\rfloor}{4} \right\rfloor \leq \left\lfloor \frac{4h(n)-1}{4} \right\rfloor < h(n).$$

And, from Lemma 11.2, we have, that $h(n+1) > N$. □

Theorem 11.1 does not allow to find an asymptotically better lower bound on $f(n)$ than the trivial lower bound $\lceil (\frac{4}{3})^n \rceil$. Note that using Lemma 11.1 one proves easily that the limit

$$\beta = \lim_{n \to \infty} \frac{\ln f(n)}{n}$$

exists. This limit satisfies $\frac{4}{3} \leq e^\beta \leq \sqrt[3]{4}$. The upper bound following from Lemma 11.1 and $f(3) = 4$. The determination of β is open.

Proposition 11.1. *[Dutour Sikirić, Itoh and Poyarkov (2007)] One has $h(4) = 7$.*

Proof. The following set of center coordinates proves that $h(4) \leq 7$.

$$\{(0,0,0,0), (1,1,1,1), (2,2,2,2), (3,3,3,3), (0,0,1,1), (1,1,2,2), (2,2,3,3)\}.$$

From Theorem 11.1, we have $h(4) \geq 6$. Assume that $h(4) = 6$ and take a blocking set of six 2-cubes with centers $\bar{a}^i = (a_1^i, a_2^i, a_3^i, a_4^i)$, $1 \leq i \leq 6$.

If three vectors a^{i_1}, a^{i_2}, a^{i_3} have equal coordinate j, then by a reasoning similar to Lemma 11.2, one finds a vector which does not overlap with those vectors.

So, the above situation does not occur and for every coordinate j, there exist two pairs $\{a^{i_1}, a^{i_2}\}$, $\{a^{i_3}, a^{i_4}\}$, which have equal j coordinates.

We have two pairs A and B in the first column. Take a pair A' in second column and assume that it does not intersect with A. Denote by P' the set of vectors obtained by removing the vector corresponding to the sets A and A' and the first and second coordinates of the remaining vectors. P' is a set of two vectors in dimension 2; hence, it is not blocking. So, we can find a 2-cube, which does not overlap with P. So, any of six pairs from three other columns must intersect with A and B.

But we have only 4 different ways to intersect A and B. So, two pairs from column $2 - 4$ are equal. But, if two pairs are equal, then they do not intersect, which is impossible. So, $h(4) > 6$. $\qquad\square$

Theorem 11.1 and Proposition 11.1 imply the following inequalities:

$$f(5) \geq h(5) \geq 10 \quad \text{and} \quad f(6) \geq h(6) \geq 14.$$

By running extensive random computations we found more than $140,000$ non-extensible cube packings in dimension 5 with 12 cubes; they belong to 203 orbits. Hence, it seems reasonable for us to conjecture that in fact $f(5) = 12$ and that the number of orbits of non-extensible cube packings with 12 cubes is "small", i.e. a few hundred.

But dimension 6 is already very different. We know that $f(6) \leq 16$ but we are unable to find by random methods a single non-extensible cube packing with less than 20 cubes.

We now consider cube packing with high density.

Take a cube packing of $\mathbb{R}^n$ with center set $\{x^k\}$, $1 \leq k \leq N$. Select a coordinate i and an index j and form a cube packing of $\mathbb{R}^{n-1}$, called *induced cube packing on layer j*, by selecting all x^k with $x_i^k = j, j + 1 \pmod 4$ and then creating the vector $(x_1^k, \ldots, x_{i-1}^k, x_{i+1}^k, \ldots, x_n^k)$.

Lemma 11.3. *[Dutour Sikirić, Itoh and Poyarkov (2007)] If $\mathcal{CP}$ is a cube packing with $2^n - \delta$ cubes, then its induced cube packings have at least $2^{n-1} - \delta$ cubes.*

Proof. Select a coordinate and denote by n_j the number of 2-cubes of $\mathcal{CP}$, with $x_i = j$. One has $n_0 + n_1 + n_2 + n_3 = 2^n - \delta$.

The number of 2-cubes of the induced cube packing on layer j is $y_j = n_j + n_{j+1}$. One writes $y_j = 2^{n-1} - \delta_j$ with $\delta_j \geq 0$, since the induced cube packing is a packing. Clearly, one has $\delta_0 + \delta_1 + \delta_2 + \delta_3 = 2\delta$.

We have $n_j + n_{j+1} = 2^{n-1} - \delta_j$; so, one gets, by subtracting $n_j - n_{j+2} = \delta_{j+1} - \delta_j$, which implies:

$$\delta_0 - \delta_1 + \delta_2 - \delta_3 = 0.$$

Every vector $\Delta = (\delta_0, \delta_1, \delta_2, \delta_3) \in \mathbb{Z}_+^4$, satisfying the above relation, can be expressed in the form $c_0(1,0,0,1) + c_1(1,1,0,0) + c_2(0,1,1,0) + c_3(0,0,1,1)$ with $c_j \in \mathbb{Z}_+$. This implies $\delta_j = c_j + c_{j+1} \leq \sum c_j = \delta$. $\square$

Theorem 11.2. *[Dutour Sikirić, Itoh and Poyarkov (2007)] In dimension n, every cube packing with $2^n - \delta$ cubes for $\delta = 1, 2, 3$ can be extended to a cube tiling.*

Proof. The proof is by induction on n. The case $n = 3$ can be solved, for example, by computer. Take $n \geq 4$ and a cube packing $\mathcal{CP}$ with $2^n - \delta$ cubes and denote by $hole(\mathcal{CP})$ its hole in $\mathbb{R}^n$. Let us consider the layering along the coordinate i. By Lemma 11.3, the induced cube packings have $2^{n-1} - \delta_j$ cubes with $\delta_j \leq 3$. So, one can complete them to form a cube packing of $\mathbb{R}^{d-1}$. Denote by $\mathcal{CC}_i = [0,2)^{i-1} \times [0,1) \times [0,2)^{n-i}$ the half of a 2-cube cut along the coordinate i. The induced cube packings are extensible by the induction hypothesis. This means that $hole(\mathcal{CP})$ is the union of 2δ cut cubes $\mathcal{CC}_i$. Denote by $\Delta_i = (\delta_0, \delta_1, \delta_2, \delta_3)$ the corresponding vector; by the analysis of Lemma 11.3 $\Delta_i = c_0(1,0,0,1) + c_1(1,1,0,0) + c_2(0,1,1,0) + c_3(0,0,1,1)$ for some $c_i \in \mathbb{Z}_+$ with $\sum c_j = \delta$.

Suppose that for a given i, the vector Δ_i contains the pattern $(0,1)$. This means that on 'one layer we have exactly one translate, say $v + \mathcal{CC}_i$, of $\mathcal{CC}_i$. Select any other coordinate i', $v + \mathcal{CC}_i$ is split into two parts, say $v^1 + \mathcal{CC}_{i'}$ and $v^2 + \mathcal{CC}_{i'}$ by the layers along the coordinate i'. Since, an adjacent layer is completely filled, this means that $v^2 = v^1 \pm e_{i'}$. Hence, they form a cube and the cube packing is extensible.

If $\delta = 1$, then up to isomorphism, one has $\Delta_i = (0,1,1,0)$. So, the above consideration proves that the cube packing is extensible.

Suppose that for a given coordinate i, $\Delta_i = (x, x, 0, 0)$ with $x = 2$ or 3. The 0-th layer is filled with x translates of set $\mathcal{CC}_i$. Take another coordinate, say i', and consider the partition of $hole(\mathcal{CP})$ into translates of $\mathcal{CC}_{i'}$. By intersecting with the 0-th layer, one obtains $2x$ intersections. But since the third layer is full, it is necessary for the translate of $\mathcal{CC}_{i'}$ to overlap only on 1-st layer. This means that they make a cube tiling.

If $\delta = 2$, then the vector Δ_i takes, up to isomorphism, one of three different forms: $(2,2,0,0)$, $(1,2,1,0)$ or $(1,1,1,1)$. First two cases have been proved to be extensible.

Now assume that for a given coordinate i, one has $\Delta_i = (1,1,1,1)$. Assume also that the cube packing is non-extensible. Take one translate $v + \mathcal{CC}_i$ on layer j in $hole(\mathcal{CP})$. It is split into two parts by the translates of $\mathcal{CC}_{i'}$. Since we assume that the cube packing is non-extensible, one of these translates overlaps on layer $j - 1$ and the other one on layer $j + 1$. One obtains a unique stair structure as illustrated below in a two-dimensional section:

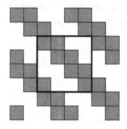

Now select another coordinate i'' (since $n \geq 4$) and see that $hole(\mathcal{CP})$ cannot be decomposed into translates of $\mathcal{CC}_{i''}$. So, if $\delta = 2$, then all cube packings are extensible.

If $\delta = 3$, then for a given coordinate i, one has clearly, up to isomorphism, $\Delta_i = (3,3,0,0)$, $(2,1,1,2)$ or $(2,3,1,0)$. The cases $(3,3,0,0)$ and $(2,3,1,0)$ are extensible by the above analysis. Let us consider the case $(2,1,1,2)$ and assume that the cube packing is non-extensible. The 1-st and 2-nd layers consist of only one translate of $\mathcal{CC}_i$, which we write as $v^1 + \mathcal{CC}_i$ and $v^2 + \mathcal{CC}_i$. The translate $v^2 + \mathcal{CC}_i$ is split into two by the translate of $\mathcal{CC}_{i'}$ appearing in the decomposition of $hole(\mathcal{CP})$ along coordinate i', i.e. $v^2 + \mathcal{CC}_i \subset w^1 + \mathcal{CC}_{i'} \cup w^2 + \mathcal{CC}_{i'}$. If $w_i^1 = w_i^2$, then one has a cube, which is excluded. So, $w_i^1 \neq w_i^2$. This implies that $v^2 = v^1 + e_i \pm e_{i'}$. But this is impossible, since i' is arbitrary. So, the cube packing is as expected extensible. $\quad\square$

Given an n-dimensional non-extensible cube packing with $2^n - \delta$, its *lifting* is a $(n + 1)$-dimensional non-extensible cube packing obtained by adding a layer of cube tiling; the iteration of lifting is also called lifting.

Conjecture 11.1. *Take $\mathcal{CP}$ a non-extensible d-dimensional cube packing with $2^n - \delta$ cubes. On its hole we conjecture:*

(1) If $\delta = 4$ then $hole(\mathcal{CP})$ is obtained as the hole of the lifting of the unique

non-extensible cube packing in dimension 3.

(2) The case $\delta = 5$ does not occur.

(3) If $\delta = 6$, then hole($\mathcal{CP}$) is obtained as the hole of the lifting of one of two non-extensible cube packings in dimension 4.

(4) If $\delta = 7$, then hole($\mathcal{CP}$) is obtained as the hole of the lifting of a non-extensible cube packing in dimension 4.

This conjecture is supported by extensive numerical computations. We can obtain an infinity of non-extensible cube packings with $2^n - 8$ cubes by doing layering of two $(n - 1)$-dimensional non-extensible cube packings with $2^{n-1} - 4$ cubes. This phenomenon does not occur for non-extensible cube packings with $2^n - 9$ cubes, but we are not able to state a reasonable conjecture for this case.

11.4   The second moment

Given a cube packing $\mathcal{CP}$ and $z \in \mathbb{Z}^n$, $N_z(\mathcal{CP})$ is defined as the number of 2-cubes of $\mathcal{CP}$ contained in $z + [0, 4)^n$.

This function is usefully illustrated in Figure 11.5.

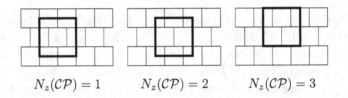

$$N_z(\mathcal{CP}) = 1 \qquad N_z(\mathcal{CP}) = 2 \qquad N_z(\mathcal{CP}) = 3$$

Fig. 11.5   The function N_z for a 2-dimensional tiling

Given a $4\mathbb{Z}^n$-periodic function f, its average is

$$E(f) = \frac{1}{4^n} \sum_{z \in \{0,1,2,3\}^n} f(z).$$

We denote $m_i(\mathcal{CP})$ the i-moment of $\mathcal{CP}$, i.e. the average of $N_z^i(\mathcal{CP})$.

Theorem 11.3. *[Dutour Sikirić, Itoh and Poyarkov (2007)] Let $\mathcal{CP}$ be a cube packing with N cubes. One has:*

$$m_1(\mathcal{CP}) = \left(\frac{3^n}{4^n}\right) N$$

and

$$m_1(\mathcal{CP}) + N(N-1)2^{-n} + 2^{-n}n\{2q(q-1) + rq\} \le m_2(\mathcal{CP})$$

with $N = 4q + r$ *and* $0 \le r \le 3$.

Proof. Take N 2-cubes $A^1, \ldots, A^N$ with centers $a^1, \ldots, a^N$. The 4-cube $a + [0,4)^n$ with corner $(a_1, \ldots, a_n)$ contains the 2-cube with center $b = (b_1, \ldots, b_n)$ if and only if $a_i \ne b_i$ for every i. Take all 4-cubes $C_1, \ldots, C_{4^n}$.

Every 2-cube A^i is contained in 3^n 4-cubes C_k. Denote by n_j the number of 2-cubes A^i, contained in the 4-cube C_j. By definition, the first moment has the expression:

$$m_1(\mathcal{CP}) = \frac{1}{4^n} \sum_k n_k = \frac{1}{4^n}\left(3^n N\right).$$

The second moment is equal to $m_2(\mathcal{CP}) = \frac{1}{4^n}\sum_k n_k^2$. Let t_{ij} be the numbers of 4-cubes containing the 2-cubes A^i and A^j. One has the relation:

$$\sum_{1 \le i < j \le N} t_{ij} = \sum_{k=1}^{4^n} \frac{n_k(n_k - 1)}{2}, \tag{11.1}$$

which implies $4^n m_1(\mathcal{CP}) + 2\sum t_{ij} = 4^n m_2(\mathcal{CP})$. Suppose the 4-cube $c + [0,4)^n$ contains the 2-cube of center a^i and a^j. If $a_l^i = a_l^j$ then c_l can take any value different from a_l^i, which makes three possibilities; while if $a_l^i \ne a_l^j$ then two values of c_l are possible. Hence, if one denotes by μ_{ij} the number of equal coordinates of the centers a^i and a^j, then one has:

$$t_{ij} = \left(\frac{3}{2}\right)^{\mu_{ij}} 2^n \ge 2^n + 2^{n-1}\mu_{ij}.$$

The above inequality becomes an equality for $\mu_{ij} = 0$ or 1. Summing over i and j one obtains

$$\sum_{1 \le i < j \le N} t_{ij} \ge N(N-1)2^{n-1} + 2^{n-1}\sum_{1 \le i < j \le N} \mu_{ij}. \tag{11.2}$$

Let us denote by R_l the number of equal pairs in column l. By definition, one clearly has:

$$\sum_{1 \le i < j \le N} \mu_{ij} = \sum_{l=1}^{n} R_l. \tag{11.3}$$

Let us fix a coordinate l and denote by d_u the number of entries equal to u in column l. One obviously has:

$$R_l = \sum_{u=0}^{3} \frac{d_u(d_u - 1)}{2}, \quad d_u \ge 0 \quad \text{and} \quad \sum_{u=0}^{3} d_u = N.$$

The Euclidean division $N = 4q + r$ and elementary optimization, with respect to the constraints, allow us to write:

$$R_l \geq 2q(q-1) + rq. \qquad (11.4)$$

The proof follows by combining (11.1), (11.2), (11.3) and (11.4). □

Note that the value of $m_1(\mathcal{CP})$ was already obtained in [Dolbilin, Itoh and Poyarkov (2005)]. For a fixed n and N, we do not know which cube packings minimize the second moment. However, in Theorem 11.4 we characterize the cube tilings of highest second moment.

Consider the following space of functions:

$$\mathcal{G} = \left\{ \begin{array}{c} f : \{0,1,2,3\}^n \to \mathbb{R} \\ \forall x \in \{0,1,2,3\}^n \text{ one has } \sum_{x+\{0,1\}^n} f(x) = 1 \text{ and } f(x) \geq 0 \end{array} \right\}.$$

It is easy to see that cube tilings correspond to $(0,1)$ vector in $\mathcal{G}$. Therefore, the problem of minimizing the second moment over cube tilings is an integer programming problem for a convex functional.

Theorem 11.4. *[Dutour Sikirić, Itoh and Poyarkov (2007)] The regular cube tiling is the cube tiling with highest second moment.*

Proof. Given a function $f \in \mathcal{G}$, let us define

$$M_i(f)(x) = \begin{cases} f(x) + f(x + e_i) & \text{if } x_i = 0 \text{ or } 2, \\ 0 & \text{if } x_i = 1 \text{ or } 3. \end{cases}$$

The function $M_i(f)$ belongs to $\mathcal{G}$. Geometrically $M_i(f)$ is the cube packing obtained by merging two induced cube packings on coordinate i and layer 0 and 2. We will prove $E(N_z(M_i(f))^2) \geq E(N_z(f)^2)$. Without loss of generality, one can assume, $i = 1$.

The key inequality, used in computation below, is:

$$(x_0 + x_1 + x_2)^2 + (x_1 + x_2 + x_3)^2 + (x_2 + x_3 + x_0)^2 + (x_3 + x_0 + x_1)^2$$
$$\leq 2(x_0 + x_1 + x_2 + x_3)^2 + (x_0 + x_1)^2 + (x_2 + x_3)^2 \text{ if } x_i \geq 0.$$

Define $f_{z_2}(z_1) = \sum_{u_2 \in \{0,1,2\}^{n-1}} f(z_1, z_2 + u_2)$ and obtain:

$$4^n E(N_z(M_1(f))^2)$$
$$= \sum_{z \in \{0,1,2,3\}^n} (\sum_{u \in \{0,1,2\}^n} M_1(f)(z+u))^2$$
$$= \sum_{z_1=0}^{3} \sum_{z_2 \in \{0,1,2,3\}^{n-1}} (\sum_{u_1=0}^{2} \sum_{u_2 \in \{0,1,2\}^{n-1}} M_1(f)(z_1 + u_1, z_2 + u_2))^2$$
$$= \sum_{z_2 \in \{0,1,2,3\}^{n-1}} \sum_{z_1=0}^{3} (\sum_{u_1=0}^{2} M_1(f_{z_2})(z_1 + u_1))^2$$
$$= \sum_{z_2 \in \{0,1,2,3\}^{n-1}} \{2(\sum_{u_1=0}^{3} f_{z_2}(u_1))^2 + (\sum_{u_1=0}^{1} f_{z_2}(u_1))^2$$
$$+ (\sum_{u_1=2}^{3} f_{z_2}(u_1))^2\}$$
$$\geq \sum_{z_2 \in \{0,1,2,3\}^{n-1}} \sum_{z_1=0}^{3} (\sum_{u_1=0}^{2} f_{z_2}(z_1 + u_1))^2 = 4^n E(N_z(f)^2).$$

Hence, using the operation $M_1 \ldots M_n$, we can only increase the second moment. So, one gets:

$$E(N_z(M_1 \ldots M_n(f))^2) \geq E(N_z(f)^2) \text{ for all } f \in \mathcal{G}.$$

It is easy to see that $M_1 \ldots M_n(f)$ is the function with $f(x) = 1$ if x is a $(0, 2)$ vector and 0, otherwise; hence, it is exactly the vector set of the regular cube tiling. □

Note that it is easy to see that $m_2 = (\frac{5}{2})^n$ for the regular cube tiling.

11.5 Appendix: Crystallographic groups

A *periodic point set* S in $\mathbb{R}^n$ is a set of points that is invariant under a group of translations under n independent vectors. The set S is called *discrete* if for any $x \in S$ we can find a neighborhood V of x such that $S \cap V = \{x\}$. A transformation f of $\mathbb{R}^n$ is called *affine* if it is the sum of a translation and a linear transformation. A crystallographic group is a group that can be the group of affine invertible transformations preserving a non-empty discrete periodic point set S. We impose that S is non-empty in order to rule out this special case for which the group of isometry is the full group of affine symmetries of $\mathbb{R}^n$.

In Figure 11.6 we list the 17 types of crystallographic groups in dimension 2, they are also called wallpaper groups or plane groups. In this figure all groups are realized by isometries of the plane. We represent a fundamental parallelogram for the translational symmetries. If there is some rotations, then we indicate the center by a filled polygon. This polygon has k sides with k the order of the rotation. A line, which does not belong to the fundamental parallelogram correspond to a plane symmetry, while a dotted line corresponds to a plane symmetry followed by a translation along that line.

We now explain some of the basic facts about crystallographic groups without proof. We refer to [O'Keefe and Hyde (1996); Opgenorth, Plesken and Schulz (1998); Hahn et al. (2002)] for more details. Let us take a crystallographic group G for a system of translation vectors $(v_1, \ldots, v_n)$. We can assume that $(v_1, \ldots, v_n)$ is maximal that is if any translation of vector v belongs to G then v is a combination with integral coefficients of $v_1, \ldots, v_n$. Define H to be the subgroup formed by the translations belonging to G. The key point is that the group H is a normal subgroup of G of finite index. The quotient group G/H is named the *point group* $Point(G)$ of the crystallographic group and it acts naturally on $\mathbb{R}^n$ as a linear operator. It also acts naturally on the quotient of $\mathbb{R}^n$ by the group of translations along $v_1, \ldots, v_n$. If one expresses the elements of the point group in terms of the basis $(v_1, \ldots, v_n)$ then one gets a group of integral matrices. For any finite matrix group G one can find a scalar product, which is preserved by the group, that is a scalar product for which G consists of isometries. The set of such scalar products forms a vector space, named the *Bravais space* or *crystal system*.

In dimension 2, the possible *crystal systems* are named parallelogram, rectangle, square and the lozenge with angles of $2\pi/3$ (see Figure 11.6). For each of those crystal systems corresponds a set of possible point groups. Then to each point group there is a set of possible crystallographic groups. In dimension 3 the crystal systems are named Hexagonal, Cubic, Orthorhombic, Tetragonal, Rhombohedral, Monoclinic and Triclinic. The number of possible point groups is 32. The 3-dimensional crystallographic groups ("Space groups") have been classified independently by E.S. Fedorov, A.M. Schoenflies and W. Barlow in the 19th century. There are 219 classes up to equivalence and 230 classes if one distinguishes up to reflections. They have a special naming system and we refer to the "International Tables for Crystallography" [Hahn et al. (2002)] for any question on almost all of their properties. For each space group we give its number in the classification, its Hermann Mauguin symbol and its Shoenflies symbol. Another nomenclature has been devised in [Conway, Delgado Friedrichs, Huson and Thurston (2001)]. The reader interested in the practical enumeration of space groups and their higher dimensional generalization, may consult [Opgenorth, Plesken and Schulz (1998)] (they are known up to dimension 6). A basis of this work is the enumeration of finite subgroups of $GL_n(\mathbb{Z})$ (see [Plesken and Pohst (1977); Nebe and Plesken (1995); Nebe (1996); Plesken (1985)] and references therein). For practical computation with high dimensional space groups, see [Eick and Souvignier (2006)].

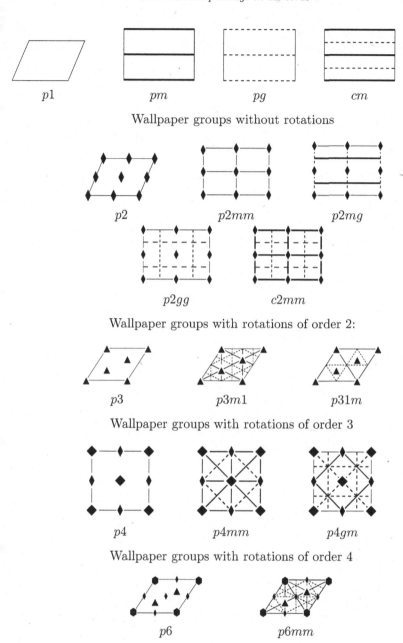

p1   pm   pg   cm

Wallpaper groups without rotations

p2   p2mm   p2mg

p2gg   c2mm

Wallpaper groups with rotations of order 2:

p3   p3m1   p31m

Wallpaper groups with rotations of order 3

p4   p4mm   p4gm

Wallpaper groups with rotations of order 4

p6   p6mm

Wallpaper groups with rotations of order 6

Fig. 11.6   The 17 wallpaper groups

Chapter 12

Continuous random cube packings in cube and torus

We consider sequential random packing of cubes $z + [0,1]^n$ with $z \in \frac{1}{N}\mathbb{Z}^n$ into the cube $[0,2]^n$ and the torus $\mathbb{R}^n/2\mathbb{Z}^n$ as $N \to \infty$. In the cube case $[0,2]^n$ as $N \to \infty$ the random cube packings thus obtained are reduced to a single cube with probability $1 - O\left(\frac{1}{N}\right)$. In the torus case the situation is different: for $n \leq 2$, sequential random cube packing yields cube tilings, but for $n \geq 3$ with strictly positive probability, one obtains non-extensible cube packings.

So, we introduce the notion of combinatorial cube packing, which instead of depending on N depend on some parameters. We use them to derive an expansion of the packing density in powers of $\frac{1}{N}$. The explicit computation is done in the cube case. In the torus case, the situation is more complicated and we restrict ourselves to the case $N \to \infty$ of strictly positive probability. We prove the following results for torus combinatorial cube packings:

- We give a general Cartesian product construction.
- We prove that the number of parameters is at least $\frac{n(n+1)}{2}$ and we conjecture it to be at most $2^n - 1$.
- We prove that cube packings with at least $2^n - 3$ cubes are extensible.
- We find the minimal number of cubes in non-extensible cube packings for n odd and $n \leq 6$.

This chapter is adapted from [Dutour Sikirić and Itoh (2010)].

12.1 Introduction

Two cubes $z + [0,1]^n$ and $z' + [0,1]^n$ are *non-overlapping* if the relative interiors $z + (0,1)^n$ and $z' + (0,1)^n$ are disjoints. A family of cubes $(z^i +$

$[0,1]^n)_{1 \leq i \leq m}$ with $z^i \in \frac{1}{N}\mathbb{Z}^n$ and $N \in \mathbb{Z}_{>0}$ is called a *discrete cube packing* if any two cubes are non-overlapping. We consider packing of cubes $z + [0,1]^n$ with $z \in \frac{1}{N}\mathbb{Z}^n$ into the cube $[0,2]^n$ and the torus $\mathbb{R}^n/2\mathbb{Z}^n$. In those two cases, two cubes $z + [0,1]^n$ and $z' + [0,1]^n$ are non-overlapping if and only if there exists an index $i \in \{1, \ldots, n\}$ such that $z_i \equiv z'_i + 1 \pmod 2$. A discrete cube packing is a *tiling* if the number of cubes is 2^n and it is *non-extensible* if it is maximal by inclusion with less than 2^n cubes.

A *sequential random cube packing* consists of putting a cube $z + [0,1]^n$ with $z \in \frac{1}{N}\mathbb{Z}^n$ uniformly at random in the cube $[0,2]^n$ or the torus $\mathbb{R}^n/2\mathbb{Z}^n$ until a maximal packing is obtained. Let us denote by $M_N^C(n)$, $M_N^T(n)$ the random variables of number of cubes of those non-extensible cube packings and by $E(M_N^C(n))$, $E(M_N^T(n))$ their expectation. We are interested in the limit $N \to \infty$ and we prove that if $N > 1$ then

$$E(M_N^U(n)) = \sum_{k=0}^{\infty} \frac{U_k(n)}{(N-1)^k} \text{ with } U \in \{C, T\} \text{ and } U_k(n) \in \mathbb{Q}. \quad (12.1)$$

In the cube case we prove that $C_k(n)$ are polynomials of degree k, which we compute for $k \leq 6$ (see Theorem 12.2). In particular, $C_0 = 1$, since as $N \to \infty$ with probability $1 - O(\frac{1}{N})$, one cannot add any more cube after the first one. In the torus case the coefficients $T_k(n)$ are no longer polynomials in n. The first coefficient $T_0(n) = \lim_{N \to \infty} E(M_N^T(n))$ is known only for $n \leq 4$ (see Table 12.1). But we prove in Theorem 12.5 that if $n \geq 3$ then $T_0(n) < 2^n$. This upper bound is related to the existence in dimension $n \geq 3$ of non-extensible torus cube packings (see Figure 12.5, Table 12.1, Theorem 12.5 and Section 12.5).

Those results are derived using the notion of *combinatorial cube packings* which is introduced in Section 12.2. A combinatorial cube packing does not depend on N but instead on some parameters t_i; to a cube or torus discrete cube packing $\mathcal{CP}$, one can associate a combinatorial cube packing $\mathcal{CP}' = \phi(\mathcal{CP})$. Given a combinatorial cube packing $\mathcal{CP}$ the probability $p(\mathcal{CP}, N)$ of obtaining a discrete cube packing $\mathcal{CP}'$ with $\phi(\mathcal{CP}') = \mathcal{CP}$ is a fractional function of N. We say that $\mathcal{CP}$ is obtained with *strictly positive probability* if the limit $\lim_{N \to \infty} p(\mathcal{CP}, N)$ is strictly positive.

In Section 12.3 the method of combinatorial cube packings is applied to the cube case and the polynomials C_k are computed for $k \leq 6$. In the torus case, the situation is more complicated and we restrict ourselves to the case of strictly positive probability, i.e. the limit case $N \to \infty$. In Section 12.4 we consider a Cartesian product construction for continuous cube packings obtained with strictly positive probability. The related lamination construction is used to derive an upper bound on $E(M_\infty^T(n))$ in Theorem 12.5.

In Section 12.5, we consider properties of non-extensible combinatorial torus cube packings. Firstly, we prove in Theorem 12.6 that combinatorial cube packings with at least $2^n - 3$ cubes are extensible to tilings. In Propositions 12.3 and 12.4, we prove that non-extensible combinatorial torus cube packings obtained with strictly positive probability have at least $\frac{n(n+1)}{2}$ parameters and that this number is attained by a combinatorial cube packing with $n + 1$ cubes if n is odd. We conjecture that the number of parameters is at most $2^n - 1$ (see Conjecture 12.1). In Proposition 12.5 we prove that in dimension 6 the minimal number of cubes in non-extensible combinatorial cube packings is 8 and that none of those cube packings is attained with strictly positive probability. In Proposition 12.6 we show that in dimension 3, 5, 7 and 9, there exist combinatorial cube tilings obtained with strictly positive probability and $\frac{n(n+1)}{2}$ parameters.

In Chapter 11, we considered the case $N = 2$ and a measure of regularity called second moment, which has no equivalent here.

12.2 Combinatorial cube packings

If $z + [0,1]^n \subset [0,2]^n$ and $z = (z_1, \ldots, z_n) \in \frac{1}{N}\mathbb{Z}^n$ then $z_i \in \{0, \frac{1}{N}, \ldots, 1\}$. Take a discrete cube packing $\mathcal{CP} = (z^i + [0,1]^n)_{1 \le i \le m}$ of $[0,2]^n$. For a given coordinate $1 \le j \le n$ we set $\phi(z_j^i) = t_{i,j}$ with $t_{i,j}$ a parameter if $0 < z_j^i < 1$ and $\phi(z_j^i) = z_j^i$ if $z_j^i = 0$ or 1. If $z^i = (z_1^i, z_2^i, \ldots, z_n^i)$ then we set $\phi(z^i) = (\phi(z_1^i), \ldots, \phi(z_n^i))$ and to $\mathcal{CP}$ we associate the combinatorial cube packing $\phi(\mathcal{CP}) = (\phi(z^i) + [0,1]^n)_{1 \le i \le m}$.

Take a torus discrete cube packing $\mathcal{CP} = (z^i + [0,1]^n)_{1 \le i \le m}$ with $z^i \in \frac{1}{N}\mathbb{Z}^n$. For a given coordinate $1 \le j \le n$ we set $\phi(z_j^i) = t_{k,j}$ if $z_j^i \equiv \frac{k}{N}$ (mod 2) and $\phi(z_j^i) = t_{k,j} + 1$ if $z_j^i \equiv \frac{k}{N} + 1$ (mod 2) with $t_{k,j}$ a parameter. Similarly, we set $\phi(z^i) = (\phi(z_1^i), \ldots, \phi(z_n^i))$ and we define $\phi(\mathcal{CP}) = (\phi(z^i) + [0,1]^n)_{1 \le i \le m}$ the associated torus combinatorial cube packing.

In the remainder of this chapter we do not use the above parameters but instead renumber them into $t_1, \ldots, t_N$. Without loss of generality, we will always assume that different coordinates have different parameters. Of course we can define combinatorial cube packing without using discrete cube packings. In the cube case, the relevant cubes are of the form $z + [0,1]^n$ with $z_i = 0$, 1 or some parameter t. In the torus case, the relevant cubes are of the form $z + [0,1]^n$ with $z_i = t$ or $t + 1$ and t a parameter. Two cubes

$z^i + [0,1]^n$ and $z^{i'} + [0,1]^n$ are non-overlapping if there exists a coordinate j such that $z_j^i \equiv z_j^{i'} + 1 \pmod 2$. In the cube case this means that $z_j^i = 0$ or 1 and $z_j^{i'} = 1 - z_j^i$. In the torus case this means that z_j^i depends on the same parameter, say t, z_j^i, $z_j^{i'} = t$ or $t + 1$ and $z_j^i \neq z_j^{i'}$. A combinatorial cube packing is then a family of such cubes with any two of them being non-overlapping. Notions of tilings and extensibility are defined as well. Moreover, a discrete cube packing is extensible if and only if its associated combinatorial cube packing is extensible. Denote by $m(\mathcal{CP})$ the number of cubes of a combinatorial cube packing $\mathcal{CP}$ and by $N(\mathcal{CP})$ its number of parameters. Denote by $Comb^C(n)$, $Comb^T(n)$, the set of combinatorial cube packings of $[0,2]^n$, respectively $\mathbb{R}^n/2\mathbb{Z}^n$.

Given two combinatorial cube packings $\mathcal{CP}$ and $\mathcal{CP}'$ (either on cube or torus), we say that $\mathcal{CP}'$ is a *subtype* of $\mathcal{CP}$ if after assigning the parameter of $\mathcal{CP}$ to 0, 1, or some parameter of $\mathcal{CP}'$, we get $\mathcal{CP}'$. So, necessarily $m(\mathcal{CP}') = m(\mathcal{CP})$ and $N(\mathcal{CP}') \leq N(\mathcal{CP})$. A combinatorial cube packing is said to be *maximal* if it is not the subtype of any other combinatorial cube packing. Necessarily, a combinatorial cube packing $\mathcal{CP}$ is a subtype of at least one maximal combinatorial cube packing $\mathcal{CP}'$.

Given a combinatorial cube packing $\mathcal{CP}$ the number of discrete cube packings $\mathcal{CP}'$ such that $\phi(\mathcal{CP}') = \mathcal{CP}$ is denoted by $Nb(\mathcal{CP}, N)$. In the cube case we have $Nb(\mathcal{CP}, N) = (N - 1)^{N(\mathcal{CP})}$. The torus case is more complex but it is still possible to write explicit formulas: denote by $N_j(\mathcal{CP})$ the number of parameters which occurs in the j-th coordinate of $\mathcal{CP}$. We then get:

$$Nb(\mathcal{CP}, N) = \Pi_{j=1}^n \Pi_{k=1}^{N_j(\mathcal{CP})}(2N - 2(k - 1)). \qquad (12.2)$$

The asymptotic order of $Nb(\mathcal{CP}, N)$ is $(2N)^{N(\mathcal{CP})}$, which shows that $Nb(\mathcal{CP}, N) > 0$ for N large enough. More specifically, $N_j(\mathcal{CP}) \leq 2^n$ so $Nb(\mathcal{CP}, N) > 0$ if $N \geq 2^n$. Note that *a priori* it is possible to have $\mathcal{CP}'$ a subtype of $\mathcal{CP}$ and $Nb(\mathcal{CP}', N) > Nb(\mathcal{CP}, N)$ for small enough N.

Denote by $f_N^T(n)$ the minimal number of cubes of non-extensible discrete torus cube packings $(z^i + [0,1]^n)_{1 \leq i \leq m}$ with $z^i \in \frac{1}{N}\mathbb{Z}^n$. Denote by $f_\infty^T(n)$ the minimal number of cubes of non-extensible combinatorial torus cube packings.

Proposition 12.1. *For $n \geq 1$ we have $\lim_{N \to \infty} f_N^T(n) = f_\infty^T(n)$.*

Proof. A discrete cube packing $\mathcal{CP}$ is extensible if and only if $\phi(\mathcal{CP})$ is extensible. Thus $f_\infty^T(n) \geq f_N^T(n)$. Take $\mathcal{CP}$ a non-extensible combinatorial torus cube packing with the minimal number of cubes. By formula (12.2) there exists N_0 such that for $N > N_0$ we have $Nb(\mathcal{CP}, N) > 0$. The discrete cube packings $\mathcal{CP}'$ with $\phi(\mathcal{CP}') = \mathcal{CP}$ are non-extensible. So, we have $\lim_{N\to\infty} f_N^T(n) = f_\infty^T(n)$. $\qquad\square$

In the cube case we have for $N \geq 2$ the equality $f_N^C(n) = 1$.

Two combinatorial cube packings $\mathcal{CP}$ and $\mathcal{CP}'$ are said to be equivalent if after a renumbering of the coordinates, parameters and cubes of $\mathcal{CP}$ one gets $\mathcal{CP}'$. The automorphism group of a combinatorial cube packing is the group of equivalences of $\mathcal{CP}$ preserving it. Testing equivalences and computing stabilizers can be done using the program **nauty** [MacKay (2008)], which is a graph theory program for testing whether two graphs are isomorphic or not and computing the automorphism group (see Appendix A.1). The method is to associate to a given combinatorial cube packing $\mathcal{CP}$ a graph $Gr(\mathcal{CP})$, which characterize isomorphisms and automorphisms.

In this graph $Gr(\mathcal{CP})$ every cube of center $v^i = (v_1^i, \ldots, v_n^i)$ correspond to n vertices v_j^i and every parameter t_i correspond to two vertices t_i, $t_i + 1$. We have the following edges:

- Every v_j^i is adjacent to all $v_j^{i'}$ and to all $v_{j'}^i$.
- The vertices t_i and $t_i + 1$ are adjacent.
- If v_j^i is t_i, then we make it adjacent to the vertex t_i.

For the non-extensible cube packing of dimension 3 with 4 cubes:

$$
\begin{aligned}
c^1 &= (\quad t_1, \quad t_2, \quad t_3 \quad), \\
c^2 &= (\ t_1 + 1, \quad t_4, \quad t_5 \quad), \\
c^3 &= (\quad t_6, \quad t_2 + 1, t_5 + 1\ ), \\
c^4 &= (\ t_6 + 1, \ t_4 + 1, \ t_3 + 1\ )
\end{aligned}
\tag{12.3}
$$

the corresponding graph is shown in Figure 12.1.

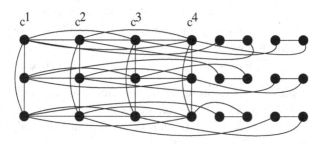

Fig. 12.1   The graph corresponding to the non-extensible continuous cube packing in dimension 3 with coordinates given in Equation (12.3)

The symmetry group of the structure is $\mathrm{Sym}(4)$, the group on the cubes is $\mathrm{Sym}(4)$, the group on the coordinates is $\mathrm{Sym}(3)$ and the group on the parameter is $\mathrm{Sym}(4)$ acting on 6 points. The corresponding programs are available from [Dutour (2008)].

We now explain the sequential random cube packing. Given a discrete cube packing $\mathcal{CP} = (z^i + [0,1]^n)_{1 \leq i \leq m}$ denote by $Poss(\mathcal{CP})$ the set of cubes $z + [0,1]^n$ with $z \in \frac{1}{N}\mathbb{Z}^n$ which do not overlap with $\mathcal{CP}$. Every possible cube $z + [0,1]^n$ is selected with equal probability $\frac{1}{|Poss(\mathcal{CP})|}$. The sequential random cube packing process is thus a process that adds cubes until the discrete cube packing is non-extensible or is a tiling.

Fix a combinatorial cube packing $\mathcal{CP}$, $N \geq 2^n$ and a discrete cube packing $\mathcal{CP}'$ such that $\phi(\mathcal{CP}') = \mathcal{CP}$. To any cube $w + [0,1]^n \in Poss(\mathcal{CP}')$ we associate the combinatorial cube packing $\mathcal{CP}_w = \phi(\mathcal{CP}' \cup \{w + [0,1]^n\})$. The set $Poss(\mathcal{CP}')$ is partitioned into classes $Cl_1, \ldots, Cl_r$ with two cubes $w + [0,1]^n$ and $w' + [0,1]^n$ in the same class if $\mathcal{CP}_w = \mathcal{CP}_{w'}$. The combinatorial cube packing associated to Cl_i is denoted by $\mathcal{CP}_i$. The set $\{\mathcal{CP}_1, \ldots, \mathcal{CP}_r\}$ of classes depends only on $\mathcal{CP}$. If we had chosen some $N \leq 2^n$, then some of the preceding $\mathcal{CP}_i$ might not have occurred. So, we have

$$|Cl_i(N)| = \frac{Nb(\mathcal{CP}_i, N)}{Nb(\mathcal{CP}, N)}$$

and we can define the probability $p(\mathcal{CP}, \mathcal{CP}_i, N)$ of obtaining a discrete cube packing of combinatorial type $\mathcal{CP}_i$ from a discrete cube packing of combinatorial type $\mathcal{CP}$:

$$p(\mathcal{CP}, \mathcal{CP}_i, N) = \frac{|Cl_i(N)|}{|Cl_1(N)| + \cdots + |Cl_r(N)|}$$
$$= \frac{Nb(\mathcal{CP}_i, N)}{Nb(\mathcal{CP}_1, N) + \cdots + Nb(\mathcal{CP}_r, N)}.$$

Given a combinatorial cube packing $\mathcal{CP}$ with m cubes a *path* $p = \{\mathcal{CP}^0, \mathcal{CP}^1, \ldots, \mathcal{CP}^m\}$ is a way of obtaining $\mathcal{CP}$ by adding one cube at a time starting from $\mathcal{CP}^0 = \emptyset$ and ending at $\mathcal{CP}^m = \mathcal{CP}$. The probability to obtain $\mathcal{CP}$ along a path p is

$$p(\mathcal{CP}, p, N) = p(\mathcal{CP}^0, \mathcal{CP}^1, N) \times \cdots \times p(\mathcal{CP}^{m-1}, \mathcal{CP}^m, N).$$

The probability $p(\mathcal{CP}, N)$ to obtain $\mathcal{CP}$ is the sum over all the paths p leading to $\mathcal{CP}$ of $p(\mathcal{CP}, p, N)$. The probabilities $p(\mathcal{CP}, p, N)$ and $p(\mathcal{CP}, N)$ are fractional functions of N, which implies that the limit values $p(\mathcal{CP}, \infty)$, $p(\mathcal{CP}, p, \infty)$ and $p(\mathcal{CP}, \mathcal{CP}', \infty)$ are well defined.

As N goes to ∞ we have the asymptotic behavior

$$|Cl_i(N)| \simeq (2N)^{nb_i}$$

with $nb_i = N(\mathcal{CP}_i) - N(\mathcal{CP})$ the number of new parameters in $\mathcal{CP}_i$ as compared with $\mathcal{CP}$. Clearly as N goes to ∞ only the classes with the largest nb_i have $p(\mathcal{CP}, \mathcal{CP}_i, \infty) > 0$. If Cl_i is such a class then we get

$$p(\mathcal{CP}, \mathcal{CP}_i, \infty) = \frac{1}{r'}$$

with r' the number of classes Cl_i having the largest nb_i and otherwise $p(\mathcal{CP}, \mathcal{CP}_i, \infty) = 0$. Analogously, for a path p leading to $\mathcal{CP}$ we can define $p(\mathcal{CP}, p, \infty)$ and $p(\mathcal{CP}, \infty)$. We say that a combinatorial cube packing $\mathcal{CP}$ is obtained with *strictly positive probability* if $p(\mathcal{CP}, \infty) > 0$ that is for at least one path p we have $p(\mathcal{CP}, p, \infty) > 0$. For a path $p = \{\mathcal{CP}^0, \mathcal{CP}^1, \ldots, \mathcal{CP}^m\}$ we have $p(\mathcal{CP}, p, \infty) > 0$ if and only if every $\mathcal{CP}^i$ has $N(\mathcal{CP}^i)$ maximal among all possible extensions from $\mathcal{CP}^{i-1}$. This implies that each $\mathcal{CP}^i$ is maximal, i.e. is not the subtype of another type. As a consequence, we can define a sequential random cube packing process for combinatorial cube packing $\mathcal{CP}$ obtained with strictly positive probability and compute their probability $p(\mathcal{CP}, \infty)$.

A combinatorial cube packing $\mathcal{CP}$ is said to have order $k = \operatorname{ord}(\mathcal{CP})$ if $p(\mathcal{CP}, N) = \frac{1}{(N-1)^k} f(N)$ with $\lim_{N \to \infty} f(N) \in \mathbb{R}_+^*$. A combinatorial cube packing is of order 0 if and only if it is obtained with strictly positive probability.

Let us denote by $M_N^C(n)$, $M_N^T(n)$ the random variables of number of cubes of those non-extensible cube packings and by $E(M_N^C(n))$, $E(M_N^T(n))$ their expectation. From the preceding discussion we have

$$E(M_N^U(n)) = \sum_{\mathcal{CP} \in Comb^U(n)} p(\mathcal{CP}, N) m(\mathcal{CP}) \text{ with } U \in \{C, T\}.$$

Denote by $f_{>0,\infty}^T(n)$ the minimal number of cubes of non-extensible combinatorial torus cube packings obtained with strictly positive probability.

In dimension 2 (see Figure 12.2), there are three combinatorial cube tilings. One of them is attained with probability 0; it is a subtype of the remaining two which are equivalent and attained with probability $\frac{1}{2}$.

In dimension 2 the process works as follows: Put a cube $z + [0, 1]^2$ in $\mathbb{R}^2/2\mathbb{Z}^2$, $z = (t_1, t_2)$. In putting the next cube, there are two possibilities: $(t_1 + 1, t_3)$ and $(t_3, t_2 + 1)$. The two corresponding packings are shown in Figure 12.4 and they are equivalent. Continuing the process, up to equivalence, one obtains with probability 1 the cube tiling of Figure 12.3.

In dimension 3 the situation is more complicated but still manageable. At first step, one puts the vector $c^1 = (t_1, t_2, t_3)$. At second step, up to

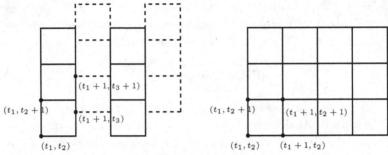

A combinatorial cube tiling obtained with probability $\frac{1}{2}$.

A combinatorial cube tiling obtained with probability 0.

Fig. 12.2   Two 2-dimensional torus combinatorial cube tilings

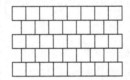

Fig. 12.3   The two dimensional cube tiling

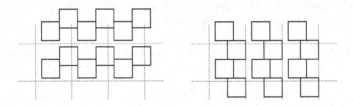

Fig. 12.4   The two possibilities for extension

equivalence, $c^2 = (t_1 + 1, t_4, t_5)$. At third step, one generates six possibilities, all with equal probability:

$$\begin{cases} (t_1 + 1, t_4 + 1, t_6), & (t_1, t_2 + 1, t_6), & (t_1, t_6, t_3 + 1), \\ (t_1 + 1, t_6, t_5 + 1), & (t_6, t_2 + 1, t_5 + 1), & (t_6, t_4 + 1, t_3 + 1). \end{cases}$$

Up to equivalence, those possibilities split into 2 cases:

- $\{(t_1, t_2, t_3), (t_1 + 1, t_4, t_5), (t_1, t_6, t_3 + 1)\}$ with probability $\frac{2}{3}$ (which are laminated),

- $\{(t_1, t_2, t_3), (t_1 + 1, t_4, t_5), (t_6, t_2 + 1, t_5 + 1)\}$ with probability $\frac{1}{3}$ (which are not laminated).

Possible extensions of the laminated case $\{(t_1, t_2, t_3), (t_1 + 1, t_4, t_5), (t_1, t_6, t_3 + 1)\}$ with probability $\frac{2}{3}$ are:

- $(t_1 + 1, t_7, t_5 + 1)$ with 1 parameter,
- $(t_1 + 1, t_4 + 1, t_7)$ with 1 parameter,
- $(t_1, t_2 + 1, t_3)$ with 0 parameter,
- $(t_1, t_6 + 1, t_3 + 1)$ with 0 parameter.

Cases with 0 parameters have probability 0, so can be neglected. So, up to equivalence, one obtains

- $\{(t_1, t_2, t_3), (t_1 + 1, t_4, t_5), (t_1, t_6, t_3 + 1), (t_1 + 1, t_7, t_5 + 1)\}$ with probability $\frac{1}{3}$,
- $\{(t_1, t_2, t_3), (t_1 + 1, t_4, t_5), (t_1, t_6, t_3 + 1), (t_1 + 1, t_4 + 1, t_7)\}$ with probability $\frac{1}{3}$,

which are extensible to tiling in the torus.

Possible extensions of the non-laminated case $\{(t_1, t_2, t_3), (t_1 + 1, t_4, t_5), (t_6, t_2 + 1, t_5 + 1)\}$ with probability $\frac{1}{3}$ are:

$$\begin{cases} (t_6 + 1, t_4 + 1, t_3 + 1), \ (t_1 + 1, t_2, t_5 + 1), \ (t_1, t_2 + 1, t_5), \\ (t_6 + 1, t_2 + 1, t_5 + 1), \ (t_1 + 1, t_4 + 1, t_5), \ (t_1, t_2, t_3 + 1). \end{cases}$$

All those choices have 0 parameter. Those possibilities are in two groups:

- $\{(t_1, t_2, t_3), (t_1 + 1, t_4, t_5), (t_6, t_2 + 1, t_5 + 1), (t_6 + 1, t_4 + 1, t_3 + 1)\}$ with probability $\frac{1}{18}$ (which are not extensible to tiling),
- 5 other cases with probability $\frac{5}{18}$ (which are extensible to tiling).

Hence the expectation of the packing density is $\frac{35}{36}$. The results are presented in Figure 12.5.

The non-extensible cube packing shown in this figure already occurs in [Lagarias, Reeds and Wang (2000)] and Chapter 11. In dimension 4, the same enumeration method works (see Table 12.1) but dimension 5 is computationally too difficult to enumerate.

Remark 12.1. A parallelotope P is an n-dimensional polytope such that there exists a lattice $L = \mathbb{Z}v_1 + \cdots + \mathbb{Z}v_n$ with $P + v$ for $v \in L$ tiling $\mathbb{R}^n$. In dimension 1 the only type of parallelotope is the interval $[0, l]$ for some $l > 0$ and the types of 2-dimensional parallelotopes are the parallelogram and centrally symmetric hexagons. The list of known results is given in Table 12.2:

Table 12.1   Number of packings and tilings for the case $N = \infty$ and $N = 2$ (see Chapter 11)

	n	1	2	3	4	5
$N = \infty$	Nr cube tilings	1	1	3	32	?
	Nr non-extensible cube packings	0	0	1	31	?
	$f_{>0,\infty}^{T}(n)$	2	4	4	6	6
	$\frac{1}{2^n} E(M_{\infty}^{T}(n))$	1	1	$\frac{35}{36}$	$\frac{15258791833}{16102195200}$	?
$N = 2$	Nr cube tilings	1	2	8	744	?
	Nr cube packings	0	0	1	139	?
	$f_{2}^{T}(n)$	2	4	4	8	$10 \le f_2^T(5)$ ≤ 12

Table 12.2   Number of combinatorial type of parallelotopes in dimension 2 to 5.

Dimension	No. types	Authors
2	2 (hexagon, parallelogram)	Dirichlet (1860)
3	5	Fedorov (1885)
4	52	Delaunay, Shtogrin (1973)
5	179377	Engel (2000)

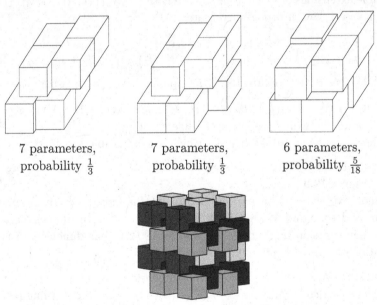

7 parameters, probability $\frac{1}{3}$ 7 parameters, probability $\frac{1}{3}$ 6 parameters, probability $\frac{5}{18}$

6 parameters, probability $\frac{1}{18}$

Fig. 12.5   The 3-dimensional combinatorial cube packings obtained with strictly positive probability; two laminations over 2-dimensional cube tilings, the rod tiling and the smallest non-extensible cube packing

The cube $[0, 1]^n$ is of course a parallelotope for the lattice $\mathbb{Z}^n$. A question of potential interest is to consider $2L$-periodic random packings of a parallelotope P. See below an example of a tiling by an hexagon

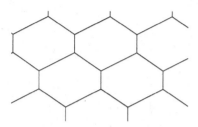 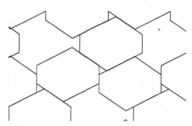

The hexagon tiling.   The $2L$ periodic tiling by the hexagon.

Clearly, translating the results obtained there to the parallelotope setting is of interest but also technically difficult.

12.3 Discrete random cube packings of the cube

We compute here the polynomials $C_k(n)$ occurring in Equation (12.1) for $k \le 6$. We compute the first three polynomials by an elementary method.

Lemma 12.1. *Put the cube $z^1 + [0, 1]^n$ in $[0, 2]^n$ and write*

$$I = \{i \ : \ z_i^1 = 0 \ or \ N\}$$

then do sequential random discrete cube packing.

(i) The minimal number of cubes in the packing is $|I| + 1$.

(ii) The expected number of cubes in the packing is $|I| + 1 + O\left(\frac{1}{N+1}\right)$.

Proof. Let us prove (i). If $|I| = 0$, then clearly one cannot insert any more cubes. We will assume $|I| > 1$ and do a reasoning by induction on $|I|$. If one puts another cube $z^2 + [0, 1]^n$, there should exist an index $i \in I$ such that $|z_i^2 - z_i^1| = 1$. Take an index $j \ne i$ such that z_j^2 is 0 or 1. The set of possibilities to add a subsequent cube is larger if $z_j \in \{0, 1\}$ than if $z_j \in \{\frac{1}{N}, \dots, \frac{N-1}{N}\}$. So, one can assume that for $j \ne i$, one has $0 < z_j^2 < 1$. This means that any cube $z + [0, 1]^n$ in subsequent insertion should satisfy $|z_i - z_i^2| = 1$, i.e. $z_i = z_i^1$. So, the sequential random cube packing can be done in one dimension less, starting with $z'^1 = (z_1, \dots, z_{i-1}, z_{i+1}, \dots, z_n)$. The induction hypothesis applies. Assertion (ii) follows easily by looking at the above process. For a given i the choice of z^2 with $0 < z_j^2 < 1$ for $j \neq i$ is the one with probability $1 - O\left(\frac{1}{N+1}\right)$. So, all neglected possibilities have probability $O\left(\frac{1}{N+1}\right)$ and with probability $1 - O\left(\frac{1}{N+1}\right)$ the number of cubes is the minimal possible, i.e. $|I| + 1$.   □

See below the 2-dimensional possibilities:

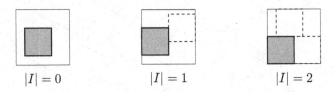

$$|I| = 0 \qquad\qquad |I| = 1 \qquad\qquad |I| = 2$$

The random variable $M_N^C(n)$ is the number of cubes in the obtained non-extensible cube-packing. $E(M_N^C(n))$ is the expected number of cubes and $E(M_N^C(n) \mid k)$ the expected number of cubes obtained by imposing the condition that the first cube $z^1 + [0,1]^n$ has $|\{i \ : \ z_i^1 = 0 \text{ or } 1\}| = k$.

Theorem 12.1. *For any $n \geq 1$, we have*

$$E(M_N^C(n)) = 1 + \frac{2n}{N+1} + \frac{4n(n-1)}{(N+1)^2} + O\left(\frac{1}{N+1}\right)^3 \quad \text{as } N \to \infty.$$

Proof. If one chooses a vector z in $\{0, \dots, N\}^n$ the probability that $|\{i \ : \ z_i^1 = 0 \text{ or } N\}| = k$ is

$$\left\{\frac{2}{N+1}\right\}^k \left\{\frac{N-1}{N+1}\right\}^{n-k} \binom{n}{k}.$$

Conditioning over $k \in \{0, 1, \dots, n\}$, one obtains

$$E(M_N^C(n)) = \sum_{k=0}^{n} \left\{\frac{2}{N+1}\right\}^k \left\{\frac{N-1}{N+1}\right\}^{n-k} \binom{n}{k} E(M_N^C(n) \mid k). \quad (12.4)$$

So, one gets

$$E(M_N^C(n)) = \left(\frac{N-1}{N+1}\right)^n E(M_N^C(n) \mid 0) + n\frac{2}{N+1}\left(\frac{N-1}{N+1}\right)^{n-1} E(M_N^C(n) \mid 1)$$
$$+ n(n-1)\frac{2}{(N+1)^2}\left(\frac{N-1}{N+1}\right)^{n-2} E(M_N^C(n) \mid 2) + O\left(\frac{1}{N+1}\right)^3.$$

Clearly $E(M_N^C(n) \mid 0) = 1$ and $E(M_N^C(n) \mid 1) = 1 + E(M_N^C(n-1))$. By Lemma 12.1, $E(M_N^C(n) \mid 2) = 3 + O\left(\frac{1}{N+1}\right)$. Then, one has $E(M_N^C(n)) =$

$1 + O\left(\frac{1}{N+1}\right)$ and

$$E(M_N^C(n)) = \left(1 - \frac{2}{N+1}\right)^n + \frac{2n}{N+1}\left(1 - \frac{2}{N+1}\right)^{n-1}\left(2 + O\left(\frac{1}{N+1}\right)\right)$$
$$+ O\left(\frac{1}{(N+1)^2}\right)$$
$$= \left\{1 - \frac{2n}{N+1} + O\left(\frac{1}{(N+1)^2}\right)\right\} + \frac{4n}{N+1} + O\left(\frac{1}{(N+1)^2}\right)$$
$$= 1 + \frac{2n}{N+1} + O\left(\frac{1}{(N+1)^2}\right).$$

Inserting this expression into $E(M_N^C(n))$ and formula (12.4) one gets the result. □

So, we get $C_0(n) = 1$, $C_1(n) = 2n$ and $C_2(n) = 4n(n-2)$. It is interesting to remark that this expression agrees with the empirical formula (9.2) by [Blaisdell and Solomon (1982)].

In order to compute $C_k(n)$ in general we use methods similar to the ones of Section 12.2. Given a cube $z + [0,1]^n$ with $z_i \in \{0, \frac{1}{N}, \ldots, 1\}$ we define a face of the cube $[0,1]^n$ in the following way: if $z_i = 0$ or 1 then we set $\psi(z_i) = 0$ or 1 whereas if $0 < z_i < 1$ we set $\psi(z_i) = t_i$ with t_i a parameter. When the parameters t_i of the vector $(\psi(z_1), \ldots, \psi(z_n))$ vary in $(0,1)$ this vector describes a face of the cube $[0,1]^n$, which we denote by $\psi(z)$. This construction was presented for the first time in [Poyarkov (2003, 2005)] and is seen in Chapter 10.

If F and F' are two faces of $[0,1]^n$, then we say that F is a sub-face of F' and write $F \subset F'$ if F is included in the closure of F'. A subcomplex of the hypercube $[0,1]^n$ is a set of faces, which contains all its sub-faces. If $\mathcal{CP}$ is a cube packing in $[0,2]^n$, then the vectors z such that $z + [0,1]^n$ is a cube which we can add to it are indexed by the faces of a subcomplex $[0,1]^n$ with the dimension giving the exponent of $(N-1)^k$. The dimension of a complex is the highest dimension of its faces. Given a discrete cube packing $\mathcal{CP}$, we have seen in Section 12.2 that the size of $Poss(\mathcal{CP})$ depends only on the combinatorial type $\phi(\mathcal{CP})$. In the cube case which we consider in this section $Poss(\mathcal{CP})$ itself depends only on the combinatorial type.

Theorem 12.2. *There exist polynomials $C_k(n)$ of n with $\deg C_k = k$ such that for any n and $N > 1$ one has:*

$$E(M_N^C(n)) = \sum_{k=0}^{\infty} \frac{C_k(n)}{(N-1)^k}.$$

The polynomials $C_k(n)$ are given in Table 12.3.

Proof.   The image $\psi(Poss(\mathcal{CP}))$ is a union of faces of $[0,1]^n$, i.e. a sub-complex of the complex $[0,1]^n$. Denote by $\dim(F)$ the dimension of a face F of the cube $[0,1]^n$. Denote by $Poss(F)$ the set of vectors $z \in \{0, \frac{1}{N}, \ldots, 1\}^n$ with $\psi(z) = F$. we have the formula:

$$|Poss(F)| = (N-1)^{\dim(F)} \quad \text{and} \quad |Poss(\mathcal{CP})| = \sum_F (N-1)^{\dim(F)}.$$

The cubes, whose corresponding face in $[0,1]^n$ have dimension $\dim(\psi(Poss(\mathcal{CP})))$ have the highest probability of being obtained. If one seeks the expansion of $E(M_N^C(n))$ up to order k and if $\mathcal{CP}$ is of order $\mathrm{ord}(\mathcal{CP})$ then we need to compute the faces of $\psi(Poss(\mathcal{CP}))$ of dimension at least $\dim(\psi(Poss(\mathcal{CP}))) - (k - \mathrm{ord}(\mathcal{CP}))$. The probabilities are then obtained in the following way:

$$p(F,N) = \frac{(N-1)^{\dim(F)}}{\sum_{F' \in S_{face}(F)} (N-1)^{\dim(F')}} \tag{12.5}$$

with

$$S_{face}(F) = \left\{ \begin{array}{c} \psi(Poss(\mathcal{CP})) \text{ with} \\ \dim(F') \geq \dim(\psi(Poss(\mathcal{CP}))) - (k - \mathrm{ord}(\mathcal{CP})) \end{array} \right\}.$$

The enumeration algorithm is then the following:

Input: Exponent k.
Output: List $\mathcal{L}$ of all inequivalent combinatorial types of non-extensible cube packings $\mathcal{CP}$ with order at most k and their probabilities $p(\mathcal{CP}, N)$ with an error of $O\left(\frac{1}{(N+1)^{k+1}}\right)$.

$\mathcal{T} \leftarrow \{\emptyset\}$.
$\mathcal{L} \leftarrow \emptyset$
while there is a $\mathcal{CP} \in \mathcal{T}$ **do**
    $\mathcal{T} \leftarrow \mathcal{T} \setminus \{\mathcal{CP}\}$
    $\psi(Poss(\mathcal{CP})) \leftarrow$ the complex of all possibilities of adding a cube to $\mathcal{CP}$
    $\mathcal{F} \leftarrow$ the faces of $\psi(Poss(\mathcal{CP}))$ of dimension at least
        $\dim(\psi(Poss(\mathcal{CP}))) - (k - \mathrm{ord}(\mathcal{CP}))$
    if $\mathcal{F} = \emptyset$ **then**
        if $\mathcal{CP}$ is equivalent to a $\mathcal{CP}'$ in $\mathcal{L}$ **then**
            $p(\mathcal{CP}', N) \leftarrow p(\mathcal{CP}', N) + p(\mathcal{CP}, N)$
        else
            $\mathcal{L} \leftarrow \mathcal{L} \cup \{\mathcal{CP}\}$
        end if
    else

```
for C ∈ F do
    CP_new ← CP ∪ {C}
    p(CP_new, N) ← p(CP, N)p(C, N)
    if CP_new is equivalent to a CP' in T then
        p(CP', N) ← p(CP', N) + p(CP_new, N)
    else
        T ← T ∪ {CP_new}
    end if
end for
end if
end while
```

Let us prove that the coefficients $C_k(n)$ are polynomials in the dimension n. If C is the cube $[0,1]^n$ then the number of faces of codimension l is $2^l \binom{n}{l}$, i.e. a polynomial in n of degree l. Suppose that a cube packing $CP = (z^i + [0,1]^n)_{1 \leq i \leq m}$ has $0 < z_j^i < 1$ for $n' \leq j \leq n$. Then all faces F of $\psi(Poss(CP))$ of maximal dimension $d = \dim(\psi(Poss(CP)))$ have $0 < z_j < 1$ for $n' \leq j \leq n$ and $z \in F$. When one chooses a subface of F of dimension $d - l$, we have to choose some coordinates j to be equal to 0 or 1. Denote by l' the number of such coordinates j with $n' \leq j \leq n$. There are $2^{l'} \binom{n+1-n'}{l'}$ choices and they are all equivalent. There are still $l - l'$ choices to be made for $j \leq n' - 1$ but this number is finite so in all cases the faces of $\psi(Poss(CP))$ of dimension at least $d - l$ can be grouped in a finite number of classes with the size of the classes depending on n polynomially. Moreover, the number of classes of dimension d is finite so the term of higher order in the denominator of Equation (12.5) is constant and the coefficients of the expansion of $p(F, N)$ are polynomial in n. □

12.4 Combinatorial torus cube packings and lamination construction

Lemma 12.2. *Let CP be a non-extensible combinatorial torus cube packing.*

(i) Every parameter t of CP occurs, which occurs as t also occurs as $t + 1$.

(ii) Let $C_1, \ldots, C_k$ be cubes of CP and C a cube which does not overlap with $CP' = CP - \{C_1, \ldots, C_k\}$. The number of parameters of C, which does not occur in CP' is at most $k - 1$.

Table 12.3   The polynomials $C_k(n)$. $|Comb_k^C|$ is the number of types of combinatorial cube packings $\mathcal{CP}$ with $\mathrm{ord}(\mathcal{CP}) \leq k$

k	$|Comb_k^C|$	$C_k(n)$
0	1	1
1	2	$2n$
2	3	$4n(n-2)$
3	7	$\frac{1}{3}\{28n^3 - 153n^2 + 149n\}$
4	18	$\frac{1}{2 \cdot 3^2 \cdot 5}\{2016n^4 - 21436n^3 + 58701n^2 - 40721n\}$
5	86	$\frac{1}{2^2 3^3 \cdot 5 \cdot 7}\{208724n^5 - 3516724n^4 + 18627854n^3$ $-35643809n^2 + 20444915n\}$
6	1980	$\frac{1}{2^8 3^{11} 5^5 7^3 11^3 \cdot 13 \cdot 17}\{19298687292242143297 03n^6$ $-46928283796201160537385n^5$ $+39737905659549633017 1955n^4$ $-14426599742910804137 70375n^3$ $+22052755559526213378 47422n^2$ $-11159113224667871432 41320n\}$

Proof.   (i) Suppose that a parameter t of $\mathcal{CP}$ occurs as t but not as $t+1$ in the coordinates of the cubes. Let $C = z + [0,1]^n$ be a cube having t in its j-th coordinate. If $C' = z' + [0,1]^n$ is a cube of $\mathcal{CP}$, then there exists a coordinate j' such that $z'_{j'} \equiv z_{j'} + 1 \pmod 2$. Necessarily $j' \neq j$ since $t+1$ does not occur, so $C + e_j$ does not overlap with C' as well and obviously $C + e_j$ does not overlap with C.

(ii) Let $C = z + [0,1]^n$ be a cube which does not overlap with the cubes of $\mathcal{CP}'$. Suppose that z has k coordinates $i_1 < \cdots < i_k$ such that their parameters $t_1, \ldots, t_k$ do not occur in $\mathcal{CP}'$. If $C_j = z^j + [0,1]^n$, then we fix $z_{i_j} \equiv z^j_{i_j} + 1 \pmod 2$ for $1 \leq j \leq k$ so that C does not overlap with $\mathcal{CP}$. This contradicts the fact that $\mathcal{CP}$ is extensible so z has at most $k-1$ parameters, which do not occur in $\mathcal{CP}'$. $\square$

Take two combinatorial torus cube packings $\mathcal{CP} = (z^i + [0,1]^n)_{1 \leq i \leq m}$ and $\mathcal{CP}' = (z'^j + [0,1]^{n'})_{1 \leq j \leq m'}$. We define m-independent copies of $\mathcal{CP}'$. For $1 \leq i \leq m$ denote by $(z'^{i,j} + [0,1]^{n'})_{1 \leq j \leq m'}$ a copy of $\mathcal{CP}'$; that is every parameter t'_k of z'^j is replaced by a parameter $t'_{i,k}$ in $z'^{i,j}$. One defines the combinatorial torus cube packing $\mathcal{CP} \ltimes \mathcal{CP}'$ by

$$(z^i, z'^{i,j}) + [0,1]^{n+n'} \text{ for } 1 \leq i \leq m \text{ and } 1 \leq j \leq m'.$$

Denote by $\mathcal{CP}_1$ the 1-dimensional combinatorial packing formed by $(t + [0,1], t+1+[0,1])$. The combinatorial cube packings $\mathcal{CP}_1 \ltimes \mathcal{CP}_1$ and $\mathcal{CP}_1 \ltimes (\mathcal{CP}_1 \ltimes \mathcal{CP}_1)$ are the ones on the left of Figures 12.2 and 12.5, respectively. Note that in general $\mathcal{CP} \ltimes \mathcal{CP}'$ is not isomorphic to $\mathcal{CP}' \ltimes \mathcal{CP}$.

Theorem 12.3. *Let $\mathcal{CP}$ and $\mathcal{CP}'$ be two combinatorial torus cube packings of dimension n and n', respectively.*

(i) $m(\mathcal{CP} \ltimes \mathcal{CP}') = m(\mathcal{CP})m(\mathcal{CP}')$ and $N(\mathcal{CP} \ltimes \mathcal{CP}') = N(\mathcal{CP}) + m(\mathcal{CP})N(\mathcal{CP}')$.

(ii) $\mathcal{CP} \ltimes \mathcal{CP}'$ is non-extensible if and only if $\mathcal{CP}$ and $\mathcal{CP}'$ are non-extensible.

(iii) If $\mathcal{CP}$ and $\mathcal{CP}'$ are obtained with strictly positive probability and $\mathcal{CP}$ is non-extensible then $\mathcal{CP} \ltimes \mathcal{CP}'$ is attained with strictly positive probability.

(iv) One has $f_\infty^T(n + m) \leq f_\infty^T(n)f_\infty^T(m)$ and $f_{>0,\infty}^T(n + m) \leq f_{>0,\infty}^T(n)f_{>0,\infty}^T(m)$.

Proof. Denote by $(z^i + [0,1]^n)_{1 \leq i \leq m}$ and by $(z'^j + [0,1]^{n'})_{1 \leq j \leq m'}$ the cubes of $\mathcal{CP}$ and $\mathcal{CP}'$ obtained in this order, i.e. first $z^1 + [0,1]^n$, then $z^2 + [0,1]^n$ and so on. Assertion (i) follows by simple counting.

If $\mathcal{CP}$, respectively $\mathcal{CP}'$ is extensible to $\mathcal{CP} \cup \{C\}$, $\mathcal{CP}' \cup \{C'\}$ then $\mathcal{CP} \ltimes \mathcal{CP}'$ is extensible to $(\mathcal{CP} \cup \{C\}) \ltimes \mathcal{CP}'$, respectively $\mathcal{CP} \ltimes (\mathcal{CP}' \cup \{C'\})$ and so extensible. Suppose now that $\mathcal{CP}$ and $\mathcal{CP}'$ are non-extensible and take a cube $z + [0,1]^{n+n'}$ with z expressed in terms of the parameters of $\mathcal{CP} \ltimes \mathcal{CP}'$. Then the cube $(z_1, \ldots, z_n) + [0,1]^n$ overlaps with one cube of $\mathcal{CP}$, say $z^i + [0,1]^n$. Also $(z_{n+1}, \ldots, z_{n+n'}) + [0,1]^{n'}$ overlaps with one cube of $\mathcal{CP}'$, say $z'^j + [0,1]^n$. So, $z + [0,1]^{n+n'}$ overlaps with the cube $(z^i, z'^{i,j}) + [0,1]^{n+n'}$ and $\mathcal{CP} \ltimes \mathcal{CP}'$ is non-extensible, establishing (ii).

A priori there is no simple relation between $p(\mathcal{CP} \ltimes \mathcal{CP}', \infty)$ and $p(\mathcal{CP}, \infty)$, $p(\mathcal{CP}', \infty)$. But we will prove that if $p(\mathcal{CP}, \infty) > 0$, $p(\mathcal{CP}', \infty) > 0$ and $\mathcal{CP}$ is not extensible then $p(\mathcal{CP} \ltimes \mathcal{CP}', \infty) > 0$. That is, to prove (iii) we have to provide one path, among possibly many, in the random sequential cube packing process to obtain $\mathcal{CP} \ltimes \mathcal{CP}'$ with strictly positive probability from some corresponding paths of $\mathcal{CP}$ and $\mathcal{CP}'$. We first prove that we can obtain the cubes $((z^i, z'^{i,1}) + [0,1]^{n+n'})_{1 \leq i \leq m}$ with strictly positive probability in this order. Suppose that we add a cube $z + [0,1]^{n+n'}$ after the cubes $(z^{i'}, z'^{i',1}) + [0,1]^{n+n'}$ with $i' < i$. If we choose a coordinate $k \in \{n + 1, \ldots, n + n'\}$ such that $z_k = (z^{i'}, z'^{i',1})_k + 1$ for some $i' < i$ then we still have to choose a coordinate for all other cubes. This is because all parameters in $(z'^{i,1})_{1 \leq i \leq m}$ are distinct. So, we do not gain anything in terms of dimension by choosing $k \in \{n + 1, \ldots, n + n'\}$ and the choice $(z^i, z'^{i,1})$ has the same or higher dimension. So, we can get the cubes $((z^i, z'^{i,1}) + [0,1]^{n+n'})_{1 \leq i \leq m}$ with strictly positive probability.

Suppose that we have the cubes $(z^i, z'^{i,j}) + [0,1]^{n+n'}$ for $1 \leq i \leq m$ and $1 \leq j \leq m'_0$. We will prove by induction that we can add the cubes

$((z^i, z'^{i,m'_0+1}) + [0,1]^{n+n'})_{1 \le i \le m}$. Denote by $n'_{m'_0} \le n' - 1$ the dimension of choices in the combinatorial torus cube packing $(z'^j + [0,1]^{n'})_{1 \le j \le m'_0}$.

Let $z + [0,1]^{n+n'}$ be a cube, which we want to add to the existing cube packing. Denote by $\mathcal{S}_z$ the set of i such that $z + [0,1]^{n+n'}$ does not overlap with $(z^i, z'^{i,j}) + [0,1]^{n+n'}$ on a coordinate $k \le n$. The fact that $z + [0,1]^{n+n'}$ does not overlap with the cubes $(z^i, z'^{i,j}) + [0,1]^{n+n'}$ fixes $n' - n'_{m'_0}$ coordinates of z. If $i \ne i'$ then the parameters in $z'^{i,j}$ and $z'^{i',j'}$ are different; this means that $(n' - n'_{m'_0})|\mathcal{S}_z|$ components of z are determined. Therefore, since $\mathcal{CP}$ is non-extensible, we can use Lemma 12.2 (ii) and so get the following estimate on the dimension D of choices:

$$D \le \{n' - (n' - n'_{m'_0})|\mathcal{S}_z|\} + \{|\mathcal{S}_z| - 1\}$$
$$\le n'_{m'_0} - (n' - n'_{m'_0} - 1)\{|\mathcal{S}_z| - 1\}$$
$$\le n'_{m'_0}.$$

We conclude that we cannot do better in terms of dimension than adding the cubes $((z^i, z'^{i,m'_0+1}) + [0,1]^{n+n'})_{1 \le i \le m}$, which we do. So, we have a path p with $p(\mathcal{CP} \ltimes \mathcal{CP}', p, \infty) > 0$ which proves that $\mathcal{CP} \ltimes \mathcal{CP}'$ is obtained with strictly positive probability.

(iv) follows immediately from (iii) and (ii). $\square$

There exist cube packings $\mathcal{CP}$, $\mathcal{CP}'$ obtained with strictly positive probability such that $p(\mathcal{CP} \ltimes \mathcal{CP}', \infty) > 0$, which shows that the hypothesis $\mathcal{CP}$ non-extensible is necessary in (iii).

The third 3-dimensional cube packings of Figure 12.5, named rod packing has the cubes $(h^i + [0,1]^3)_{1 \le i \le 8}$ with the following h^i:

$$
\begin{aligned}
h^1 &= (\quad t_1, \quad\quad t_2, \quad\quad t_3), & h^5 &= (\, t_6 + 1, \, t_2 + 1, \, t_5 + 1), \\
h^2 &= (\, t_1 + 1, \quad t_4, \quad\quad t_5), & h^6 &= (\quad t_1, \quad\quad t_2, \quad t_3 + 1), \\
h^3 &= (\quad t_6, \quad t_2 + 1, t_5 + 1), & h^7 &= (\, t_1 + 1, \quad t_2, \quad t_5 + 1), \\
h^4 &= (\, t_1 + 1, t_4 + 1, \quad t_5), & h^8 &= (\quad t_1, \quad t_2 + 1, \quad t_5).
\end{aligned}
$$

Taking 8 $(n-3)$-dimensional combinatorial torus cube tilings $(w^{i,j})_{1 \le j \le 2^{n-3}}$ with $1 \le i \le 8$, one defines an n-dimensional *rod tiling* combinatorial cube packing

$$(z^i, w^{i,j}) + [0,1]^n \quad \text{for} \quad 1 \le i \le 8 \quad \text{and} \quad 1 \le j \le 2^{n-3}.$$

Theorem 12.4. *The probability of obtaining a rod tiling is*

$$p_1^{15} \times q_{n-3}^8$$

where q_n is the probability of obtaining an n-dimensional cube-tiling and p_1^{15} is a rational function of n.

Proof. Up to equivalence, one can assume that in the random-cube packing process, one puts

$$z^1 = (h^1, w^{1,1}) = (t_1, t_2, t_3, \dots) \quad \text{and} \quad z^2 = (h^2, w^{2,1}) = (t_1 + 1, t_4, t_5, \dots).$$

Then there are $n(n-1)$ possible choices for the next cube, $2(n-1)$ of them are respecting the lamination. So, there are $(n-2)(n-1)$ choices which do not respect the lamination and their probability is $p_1^3 = \frac{n-2}{n}$. Without loss of generality, we can assume that one has

$$z^3 = (h^3, w^{3,1}) = (t_6, t_2 + 1, t_5 + 1, \dots).$$

In the next 5 stages we add cubes with $n-3$ new parameters each. We have more than one type to consider under equivalence and we need to determine the total number of possibilities in order to compute the probabilities.

For the cube $z^4 + [0,1]^n$ we should have three integers i_1, i_2, i_3 such that $z_{i_j}^4 \equiv z_{i_j}^j + 1 \pmod 2$. Necessarily, the i_j are all distinct, which gives $n(n-1)(n-2)$ possibilities. There are exactly 6 possibilities with $i_j \leq 3$. One of them corresponds to the non-extensible cube packing of Figure 12.5 on the first 3 coordinates which the other 5 have a non-zero probability of being extended to the rod tiling. When computing later probabilities, we used the automorphism group of the existing configuration and gather the possibilities of extension into orbits. At the fourth stage, the 5 possibilities split into two orbits:

(1) O_1^4: $(h^i, w^{i,1})$ for $i \in \{1,2,3,4\}$ with $p_1^4 = p_1^3 \frac{3}{n(n-1)(n-2)}$,
(2) O_2^4: $(h^i, w^{i,1})$ for $i \in \{1,2,3,7\}$ with $p_2^4 = p_1^3 \frac{2}{n(n-1)(n-2)}$; write $\Delta_2^4 = 3(n-3)(n-4) + 3(n-3) + 4$ the number of possibilities of adding a cube to the packing $((h^i, w^{i,1}) + [0,1]^n)_{i \in \{1,2,3,7\}}$.

When adding a fifth cube one finds the following cases up to equivalence:

(1) O_1^5: $(h^i, w^{i,1})$ for $i \in \{1,2,3,4,5\}$ with $p_1^5 = p_1^4 \frac{2}{2(n-1)(n-2)}$,
(2) O_2^5: $(h^i, w^{i,1})$ for $i \in \{1,2,3,4,7\}$ with $p_2^5 = p_1^4 \frac{2}{2(n-1)(n-2)} + p_2^4 \frac{3}{\Delta_2^4}$,
(3) O_3^5: $(h^i, w^{i,1})$ for $i \in \{1,2,3,7,8\}$ with $p_3^5 = p_2^4 \frac{1}{\Delta_2^4}$.

When adding a sixth cube one finds the following cases up to equivalence:

(1) O_1^6: $(h^i, w^{i,1})$ for $i \in \{1,2,3,4,5,6\}$ with $p_1^6 = p_1^5 \frac{1}{3(n-2)}$,
(2) O_2^6: $(h^i, w^{i,1})$ for $i \in \{1,2,3,4,5,7\}$ with $p_2^6 = p_1^5 \frac{2}{3(n-2)} + p_2^5 \frac{2}{n(n-2)}$,
(3) O_3^6: $(h^i, w^{i,1})$ for $i \in \{1,2,3,4,7,8\}$ with $p_3^6 = p_2^5 \frac{1}{n(n-2)} + p_3^5 \frac{3}{3(n-2)}$.

When adding a seventh cube one finds the following cases up to equivalence:

(1) O_1^7: $(h^i, w^{i,1})$ for $i \in \{1, 2, 3, 4, 5, 6, 7\}$ with $p_1^7 = p_1^6 + p_2^6 \frac{1}{n-1}$,

(2) O_2^7: $(h^i, w^{i,1})$ for $i \in \{1, 2, 3, 4, 5, 7, 8\}$ with $p_2^7 = p_2^6 \frac{1}{n-1} + p_3^6 \frac{2}{2(n-2)}$.

The combinatorial cube packing of eight cubes $((h^i, w^{i,1}) + [0, 1]^n)_{1 \leq i \leq 8}$ is then obtained with probability $p_1^8 = p_1^7 + p_2^7 \frac{1}{n-2}$.

Then we add cubes in dimension $n - 4$ following in fact the construction of Theorem 12.3. The parameters t_3, t_4 and t_6 appear only two times in the cube packing for the rods, which contain 6 cubes in total. So, when one adds cubes we have $8(n-3)$ choices respecting the cube packing, i.e. of the form $z^9 = (h^i, w^{i,2})$ with $w_j^{i,2} \equiv w_j^{i,1} \pmod 2$ for some $1 \leq j \leq n - 3$. We also have $3(n-3)(n-4)$ choices not respecting the rod tiling structure, i.e. of the form $z^9 = (k^i, w)$ with k^i being one of h^i for $1 \leq i \leq 3$ with t_3, t_4 or t_6 replaced by another parameter. But after adding a cube $(h^i, w^{i,2}) + [0, 1]^n$ with h^i containing t_3, t_4 or t_6 this phenomenon cannot occur. Below a *type* T_r^h of probability p_r^h is a packing formed by the 8 vectors $(h^i, w^{i,1})_{1 \leq i \leq 8}$ and $h - 8$ vectors of the form $(h^i, w^{i,2})$ amongst which r of the parameters t_3, t_4 or t_6 do not occur. Note that there may be several non-equivalent cube packings with the same type but this is not important since they have the same numbers of possibilities.

Adding 9th cube one gets:

(1) T_3^9, $p_1^9 = p_1^8 \frac{2(n-3)}{8(n-3)+3(n-3)(n-4)}$,

(2) T_2^9, $p_2^9 = p_1^8 \frac{6(n-3)}{8(n-3)+3(n-3)(n-4)}$.

Adding 10th cube one gets:

(1) T_3^{10}, $p_1^{10} = p_1^9 \frac{n-3}{7(n-3)+3(n-3)(n-4)}$,

(2) T_2^{10}, $p_2^{10} = p_1^9 \frac{6(n-3)}{7(n-3)+3(n-3)(n-4)} + p_2^9 \frac{3(n-3)}{7(n-3)+2(n-3)(n-4)}$,

(3) T_1^{10}, $p_3^{10} = p_2^9 \frac{4(n-3)}{7(n-3)+2(n-3)(n-4)}$.

Adding 11th cube one gets:

(1) T_2^{11}, $p_1^{11} = p_1^{10} \frac{6(n-3)}{6(n-3)+3(n-3)(n-4)} + p_2^{10} \frac{2(n-3)}{6(n-3)+2(n-3)(n-4)}$,

(2) T_1^{11}, $p_2^{11} = p_2^{10} \frac{4(n-3)}{6(n-3)+2(n-3)(n-4)} + p_3^{10} \frac{4(n-3)}{6(n-3)+(n-3)(n-4)}$,

(3) T_0^{11}, $p_3^{11} = p_3^{10} \frac{2(n-3)}{6(n-3)+2(n-3)(n-4)}$.

Adding 12th cube one gets:

(1) T_2^{12}, $p_1^{12} = p_1^{11} \frac{n-3}{5(n-3)+2(n-3)(n-4)}$,

(2) T_1^{12}, $p_2^{12} = p_1^{11} \frac{4(n-3)}{5(n-3)+2(n-3)(n-4)} + p_2^{11} \frac{3(n-3)}{5(n-3)+2(n-3)(n-4)}$,

(3) T_0^{12}, $p_3^{12} = p_2^{11} \frac{2(n-3)}{5(n-3)+2(n-3)(n-4)} + p_3^{11} \frac{5(n-3)}{5(n-3)+2(n-3)(n-4)}$.

Adding 13th cube one gets:

(1) T_1^{13}, $p_1^{13} = p_1^{12} \frac{4(n-3)}{4(n-3)+2(n-3)(n-4)} + p_2^{12} \frac{2(n-3)}{4(n-3)+(n-3)(n-4)}$,

(2) T_0^{13}, $p_2^{13} = p_2^{12} \frac{2(n-3)}{4(n-3)+(n-3)(n-4)} + p_3^{12} \frac{4(n-3)}{4(n-3)}$.

Adding 14th cube one gets:

(1) T_1^{14}, $p_1^{14} = p_1^{13} \frac{(n-3)}{3(n-3)+(n-3)(n-4)}$,

(2) T_0^{14}, $p_2^{13} = p_1^{13} \frac{2(n-3)}{3(n-3)+(n-3)(n-4)} + p_2^{13} \frac{3(n-3)}{3(n-3)}$.

Adding 15th cube one gets:

(1) T_0^{15}, $p_1^{15} = p_1^{14} \frac{2(n-3)}{2(n-3)+(n-3)(n-4)} + p_2^{14}$.

After that if we add a cube $z + [0,1]^n$, then necessarily z is of the form (h^i, w). So, we have 8 different $(n-3)$-dimensional cube packing problems show up and the probability is $p_1^{15} q_{n-3}^8$. $\qquad\square$

A combinatorial torus cube packing $\mathcal{CP}$ is called *laminated* if there exist a coordinate j and a parameter t such that for every cube $z + [0,1]^n$ of $\mathcal{CP}$ we have $z_j \equiv t \pmod 1$.

Theorem 12.5. *(i) The probability of obtaining a laminated combinatorial cube packing is $\frac{2}{n}$.*
(ii) For any $n \geq 1$, one has $E(M_\infty^T(n)) \leq 2^n(1 - \frac{2}{n}) + \frac{4}{n} E(M_\infty^T(n-1))$.
(iii) For any $n \geq 3$, $\frac{1}{2^n} E(M_\infty^T(n)) \leq 1 - \frac{2^n}{n!} \frac{1}{24}$.

Proof. Up to equivalence, we can assume that after the first two steps of the process, we have

$$z^1 = (t_1, \ldots, t_n) \quad \text{and} \quad z^2 = (t_1 + 1, t_{n+1}, \ldots, t_{2n-1}).$$

So, we consider lamination on the first coordinate. We then consider all possible cubes that can be added. Those cubes should have one coordinate differing by 1 with other vectors. This makes $n(n-1)$ possibilities. If a vector respects the lamination on the first coordinate then its first coordinate should be equal to t_1 or $t_1 + 1$. This makes $2(n-1)$ possibilities. So, the probability of having a family of cube respecting a lamination at the third step is $\frac{2}{n}$. But one sees easily that in all further steps, the choices breaking the lamination have a dimension strictly lower than the one respecting the lamination, so they do not occur and we get (i).

By separating between laminated and non-laminated combinatorial torus cube packings, bounding the number of cubes of non-laminated combinatorial torus cube packings by 2^n one obtains

$$E(M_\infty^T(n)) \leq \left(1 - \frac{2}{n}\right) \times 2^n + \frac{2}{n}(E(M_\infty^T(n-1)) + E(M_\infty^T(n-1))),$$

which is (ii). (iii) follows by induction starting from $\frac{1}{8}E(M_\infty^T(3)) = \frac{35}{36}$ (see Table 12.1). $\qquad \square$

12.5 Properties of non-extensible cube packings

Theorem 12.6. *If a combinatorial torus cube packing has at least $2^n - 3$ cubes, then it is extensible.*

Proof. Our proof closely follows the one of Theorem 11.2 but is different from it. Take $\mathcal{CP}'$ a combinatorial torus cube packing with $2^n - \alpha$ cubes, $\alpha \leq 3$. Take N such that $Nb(\mathcal{CP}, N) > 0$ and $\mathcal{CP}$ a discrete cube packing with $\phi(\mathcal{CP}) = \mathcal{CP}'$. If $\mathcal{CP}$ is extensible then $\mathcal{CP}'$ is extensible as well.

We select $\delta \in \mathbb{R}$ and denote by I_j the interval $[\delta + \frac{i}{2}, \delta + \frac{i+1}{2})$ for $0 \leq j \leq 3$. Denote by $n_{j,k}$ the number of cubes, whose k-th coordinate modulo 2 belong to I_j.

All cubes of $\mathcal{CP}$, whose k-th coordinate belongs to I_j, I_{j+1} form after removal of their k-th coordinate a cube packing of dimension $n - 1$, which we denote by $\mathcal{CP}_{j,k}$. We write $n_{j,k} + n_{j+1,k} = 2^{n-1} - d_{j,k}$ and obtain the equations

$$d_{0,k} - d_{1,k} + d_{2,k} - d_{3,k} = 0 \quad \text{and} \quad \sum_{j=0}^{3} d_{j,k} = 2\alpha.$$

We can then write the vector $d_k = (d_{0,k}, d_{1,k}, d_{2,k}, d_{3,k})$ in the following way:

$$d_k = c_1(1,1,0,0) + c_2(0,1,1,0) + c_3(0,0,1,1) + c_4(1,0,0,1)$$

with $\sum_{j=1}^{4} c_j = \alpha$ and $c_i \in \mathbb{Z}^+$. This implies $d_{j,k} = c_j + c_{j+1} \leq \sum c_j = \alpha$. This means that the $(n-1)$-dimensional cube packing $\mathcal{CP}_{j,k}$ has at least $2^{n-1} - 3$ cubes, so by an induction argument, we conclude that $\mathcal{CP}_{j,k}$ is extensible.

Suppose now that the k-th coordinate of the cubes in $\mathcal{CP}$ has values $0 < \delta_1 < \delta_2 < \cdots < \delta_M < 2$. So, the set of points in the complement of $\mathcal{CP}$, whose k-th coordinate belongs to the interval $[\delta_i, \delta_{i+1})$ with $\delta_{M+1} = \delta_1 + 2$ can be filled by translates of the parallelepiped $\text{Paral}_k(\alpha) = [0,1)^{k-1} \times [0,\alpha) \times [0,1)^{n-k}$.

Note that as δ varies, the vector d_k varies as well. Suppose that for some i, we have the k-th layer $[\delta_i, \delta_{i+1})$ being full and $[\delta_{i-1}, \delta_i)$ containing x translates with $x \leq 3$ of the parallelepiped $\text{Paral}_k(\delta_{i+1} - \delta_i)$. Then if one selects another coordinate k', all parallelepipeds $\text{Paral}_{k'}(\delta'_{i'+1} - \delta'_{i'})$ filling the hole delimited by the parallelepiped $\text{Paral}_k(\delta_{i+1} - \delta_i)$ will have the same position in the k-th coordinate. This means that they will form x cubes and that the cube packing is extensible. This argument solves the case $\alpha = 1$, because up to symmetry $d_k = (0,1,1,0)$.

If $\alpha = 2$, then the case of vector of coordinate d_k being equal to symmetry to $(0,2,2,0)$ or $(0,1,2,1)$ is also solved because we have seen that a full layer implies that we can fill the hole. We have the remaining case $(1,1,1,1)$. If the hole of this cube packing cannot be filled, then we have a structure of this form:

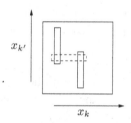

Selecting another coordinate k', we get that the two parallelepipeds $z + \text{Paral}_k(\delta_{i+1} - \delta_i)$ and $z' + \text{Paral}_k(\delta_{i'+1} - \delta_{i'})$ have $z_l = z'_l$ for $l \neq k, k'$. This is impossible if $n \geq 4$. So, if $d_k = (1,1,1,1)$ for some k and δ, then the hole can be filled.

If $\alpha = 3$, and d_k, up to symmetry, is equal to $(0,3,3,0)$ or $(0,2,3,1)$ then we have a full layer and so we can fill the hole. If the vector $d_k = (2,1,1,2)$ occurs, then by the same argument as for $(1,1,1,1)$ we can fill the hole. $\square$

Proposition 12.2. *(i) Non-extensible combinatorial torus cube packings of dimension n have at least $n + 1$ cubes.*

(ii) If $\mathcal{CP}$ is a combinatorial torus cube tiling, then in a coordinate j a parameter t occur the same number of times as t and $t + 1$.

Proof. (i) Suppose that a combinatorial torus cube packing $\mathcal{CP}$ has $m \leq n$ cubes $(z^i + [0,1]^n)_{1 \leq i \leq m}$. By fixing $z_i = z_i^i + 1$ for $i = 1, 2, \ldots, m$ we get that the cube $z + [0,1]^n$ does not overlap with $\mathcal{CP}$.

(ii) Without loss of generality, we can assume that a given parameter t occurs only in one coordinate k as t and $t+1$. The cubes occurring in the layer $[t, t+1]$, $[t+1, t+2]$ on j-th coordinate are the ones with $x_j = t$, $t+1$; we denote by V_t and V_{t+1} their volume. Now if we interchange t and $t+1$ we still obtain a tiling, so $V_t \leq V_{t+1}$ and $V_{t+1} \leq V_t$. So, $V_t = V_{t+1}$ and the number of cubes with $x_j = t$ is equal to the number of cubes with $x_j = t+1$.   □

Take a combinatorial torus cube packing CP obtained with strictly positive probability. Let us choose a path p to obtain CP. Denote by $N_{k,p}(CP)$ the number of cubes obtained with k new parameters along the path p.

Proposition 12.3. *Let CP be a non-extensible combinatorial torus cube packing, p a path with $p(CP, p, \infty) > 0$.*
 (i) $N_{n,p}(CP) = 1$ and $N_{n-1,p}(CP) = 1$.
 (ii) $N_{n-2,p}(CP) \leq 2$ and $N_{n-2,p}(CP) = 2$ if and only if CP is laminated.
 (iii) One has $N_{k,p}(CP) \geq 1$ for $0 \leq k \leq n$.
 (iv) $N(CP) = \sum_{k=0}^{n} k N_{k,p}(CP) \geq \frac{n(n+1)}{2}$.
 (v) If $N(CP) = \frac{n(n+1)}{2}$ then $N_{k,p}(CP) = 1$ for $k \geq 1$.

Proof. The first cube $z^1 + [0,1]^n$ has n new parameters, but the second cube $z^2 + [0,1]^n$ should not overlap with the first one so it has $n-1$ parameters and $N_{n,p}(CP) = 1$. Without loss of generality, we can assume that $z^1 = (t_1, \ldots, t_n)$ and $z^2 = (t_1 + 1, t_{n+1}, \ldots, t_{2n-1})$. When adding the third cube $z^3 + [0,1]^n$, we have to set up 2 coordinates depending on the parameters t_i, $i \leq 2n-1$ thus $N_{n-1,p}(CP) = 1$.

If $z_1^3 = t_1$ or $t_1 + 1$ then we have a laminated cube packing, we can add a cube with $n-2$ parameters and $N_{n-2,p}(CP) = 2$. Otherwise, we do not have a laminated cube packing, three coordinates of z^4 need to be expressed in terms of preceding cubes and thus $N_{n-2,p}(CP) = 1$.

(iii) The proof is by induction; suppose one has put $m' = \sum_{l=k}^{n} N_{l,p}(CP)$ cubes. Then the cube $z^{m'} + [0,1]^n$ has k new parameters t'_1, $\ldots$, t'_k in coordinates i_1, $\ldots$, i_k. The cube $C = z + [0,1]^n$ with $z_{i_1} = t'_1 + 1$ and $z_i = z_i^{m'}$ for $i \notin \{i_1, \ldots, i_k\}$ has $k-1$ free coordinates $\{i_2, \ldots, i_k\}$ and thus $k-1$ new parameters. So, $N_{k-1,p}(CP) \geq 1$.

(iv) and (v) are elementary.   □

Conjecture 12.1. *Let CP be a combinatorial torus cube packing and p a path with $p(CP, p, \infty) > 0$.*
 (i) For all $k \geq 1$ one has $\sum_{k=0}^{l} N_{n-k,p}(CP) \leq 2^l$.

Table 12.4   Number of non-isomorphism 1-factorization of K_{2p}

graph	|isomorphism types|	authors
K_4	1	
K_6	1	
K_8	6	[Dickson and Safford (1906)]
K_{10}	396	[Gelling and Odeh (1974)]
K_{12}	526915620	[Dinitz, Garnick and McKay (1994)]
K_{14}	1132835421602062347	[Kaski and Ostergard (2009)]

(ii) $N(\mathcal{CP}) \leq 2^n - 1$; *if* $N(\mathcal{CP}) = 2^n - 1$, *then* $\mathcal{CP}$ *is obtained via a lamination construction.*

A *perfect matching* of a graph G is a set $\mathcal{M}$ of edges such that every vertex of G belongs to exactly one edge of $\mathcal{M}$. A 1-*factorization* of a graph G is a set of perfect matchings, which partitions the edge set of G. The graph K_4 has one 1-factorization; the graph K_6 has, up to isomorphism, exactly one 1-factorization with symmetry group Sym(5), i.e. the group Sym(5) acts on 6 elements.

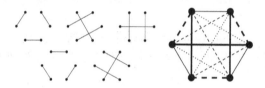

Fig. 12.6   The unique, up to isomorphism, 1-factorization of K_6

Proposition 12.4. *Let* $\mathcal{CP}$ *be a non-extensible combinatorial torus cube packing.*

(i) If n *is even then* $\mathcal{CP}$ *has at least* $n + 2$ *cubes.*

(ii) If n *is odd and* $\mathcal{CP}$ *has* $n + 1$ *cubes then* $N(\mathcal{CP}) = \frac{n(n+1)}{2}$. *Fix a coordinate* j *and a parameter* t *occurring in at least one cube. Then the number of cubes containing* t, *respectively* $t + 1$ *in coordinate* j *is exactly* 1.

(iii) If n *is odd then isomorphism classes of non-extensible combinatorial torus cube packings with* $n + 1$ *cubes are in one-to-one correspondence with isomorphism classes of 1-factorizations of* K_{n+1}.

(iv) If n *is odd then the non-extensible combinatorial torus cube packings with* $n + 1$ *cubes are obtained with strictly positive probability and* $f_\infty^T(n) = f_{>0,\infty}^T(n) = n + 1$.

Proof. We take a non-extensible cube packing $\mathcal{CP}$ with $n + 1$ cubes. Suppose that for a coordinate j we have two cubes $z^i + [0, 1]^n$ and $z^{i'} + [0, 1]^n$ with $z_j^i = z_j^{i'} = t$. If a vector z has $z_j = t + 1$, then $z + [0, 1]^n$ does not overlap with $z^i + [0, 1]^n$ and $z^{i'} + [0, 1]^n$. There are $n - 1$ remaining cubes to which $z + [0, 1]^n$ should not overlap but we have $n - 1$ remaining coordinates so it is possible to choose the coordinates of z so that $z + [0, 1]^n$ does not overlap with $\mathcal{CP}$. This is impossible, therefore parameters always appear at most 1 time as t and at most 1 time as $t + 1$ in a given coordinate.

By Lemma 12.2 every parameter t appears also as $t + 1$. So, every parameter t appears one time as t and one time as $t + 1$. This implies that we have an even number of cubes and so (i). Every coordinate has $\frac{n+1}{2}$ parameters, which gives $\frac{n(n+1)}{2}$ parameters and so (ii).

(iii) Assertion (ii) implies that any two cubes C_i and $C_{i'}$ of $\mathcal{CP}$ have exactly one coordinate on which they differ by 1. So, every coordinate corresponds to a perfect matching and the set of n coordinates to the 1-factorization.

(iv) Since parameters t appear only one time as t and $t+1$, the dimension of choices after k cubes are put is $n - k$ and one sees that such a cube packing is obtained with strictly positive probability. The existence of 1-factorization of K_{2p} (see, for example, [Alspach (2008); Harary, Robinson and Wormald (1978)]) gives $f_\infty^T(n) \leq f_{>0,\infty}^T(n) \leq n + 1$. Combined with Theorem 12.2.i, we have the result. $\square$

Conjecture 12.2. *If n is even then there exist non-extensible combinatorial torus cube packings with $n + 2$ cubes and $\frac{n(n+1)}{2}$ parameters.*

In dimension 4 there is a unique cube packing (obtained with probability $\frac{1}{480}$) satisfying this conjecture:

$$\begin{pmatrix} t_1 & t_2 & t_3 & t_4 \\ t_5 & t_6 & t_7 & t_4 + 1 \\ t_1 + 1 & t_8 & t_7 + 1 & t_9 \\ t_5 + 1 & t_8 + 1 & t_3 + 1 & t_{10} \\ t_1 + 1 & t_6 + 1 & t_7 & t_{10} + 1 \\ t_5 & t_2 + 1 & t_7 + 1 & t_9 + 1 \end{pmatrix}$$

Proposition 12.5. *(i) There are 9 isomorphism types of non-extensible combinatorial torus cube packings in dimension 6 with 8 cubes and at least 21 parameters (see Figure 12.7); they are not obtained with strictly positive probability.*

$$\begin{pmatrix}
t_1 & t_5 & t_9 & t_{14}+1 & t_{17}+1 & t_{19}\\
t_1+1 & t_6 & t_{10} & t_{13}+1 & t_{16}+1 & t_{19}\\
t_2 & t_5+1 & t_{11} & t_{13} & t_{18} & t_{20}\\
t_2+1 & t_7 & t_9+1 & t_{15} & t_{16} & t_{21}\\
t_3 & t_6+1 & t_{12} & t_{14} & t_{18}+1 & t_{21}+1\\
t_3+1 & t_8 & t_{10}+1 & t_{15}+1 & t_{17} & t_{20}+1\\
t_4 & t_7+1 & t_{12}+1 & t_{13}+1 & t_{17}+1 & t_{19}+1\\
t_4+1 & t_8+1 & t_{11}+1 & t_{14}+1 & t_{16}+1 & t_{19}+1
\end{pmatrix}
\begin{pmatrix}
t_1 & t_5 & t_9 & t_{13} & t_{17} & t_{21}\\
t_1+1 & t_6 & t_{10} & t_{14} & t_{18} & t_{21}\\
t_2 & t_5+1 & t_{10}+1 & t_{15} & t_{19} & t_{22}\\
t_2+1 & t_6+1 & t_9+1 & t_{16} & t_{20} & t_{22}\\
t_3 & t_7 & t_{11} & t_{13}+1 & t_{18}+1 & t_{22}+1\\
t_3+1 & t_8 & t_{12} & t_{14}+1 & t_{17}+1 & t_{22}+1\\
t_4 & t_7+1 & t_{12}+1 & t_{15}+1 & t_{20}+1 & t_{21}+1\\
t_4+1 & t_8+1 & t_{11}+1 & t_{16}+1 & t_{19}+1 & t_{21}+1
\end{pmatrix}$$
$$\text{21 parameters}, |Aut| = 4 \qquad\qquad \text{22 parameters}, |Aut| = 64$$

$$\begin{pmatrix}
t_1 & t_5 & t_9 & t_{13} & t_{17} & t_{21}\\
t_1+1 & t_6 & t_{10} & t_{14} & t_{18} & t_{21}\\
t_2 & t_5+1 & t_{10}+1 & t_{15} & t_{19} & t_{22}\\
t_2+1 & t_6+1 & t_9+1 & t_{16} & t_{20} & t_{22}\\
t_3 & t_7 & t_{11} & t_{13}+1 & t_{18}+1 & t_{22}+1\\
t_3+1 & t_8 & t_{12} & t_{14}+1 & t_{17}+1 & t_{22}+1\\
t_4 & t_7+1 & t_{12}+1 & t_{16}+1 & t_{19}+1 & t_{21}+1\\
t_4+1 & t_8+1 & t_{11}+1 & t_{15}+1 & t_{20}+1 & t_{21}+1
\end{pmatrix}
\begin{pmatrix}
t_1 & t_5 & t_9 & t_{13} & t_{17} & t_{21}\\
t_1+1 & t_6 & t_{10} & t_{14} & t_{18} & t_{22}\\
t_2 & t_5+1 & t_{10}+1 & t_{15} & t_{19} & t_{22}\\
t_2+1 & t_6+1 & t_9+1 & t_{16} & t_{20} & t_{22}\\
t_3 & t_7 & t_{11} & t_{13}+1 & t_{20}+1 & t_{22}+1\\
t_3+1 & t_8 & t_{12} & t_{14}+1 & t_{19}+1 & t_{21}+1\\
t_4 & t_7+1 & t_{12}+1 & t_{16}+1 & t_{17}+1 & t_{22}+1\\
t_4+1 & t_8+1 & t_{11}+1 & t_{15}+1 & t_{18}+1 & t_{21}+1
\end{pmatrix}$$
$$\text{22 parameters}, |Aut| = 64 \qquad\qquad \text{22 parameters}, |Aut| = 16$$

$$\begin{pmatrix}
t_1 & t_5 & t_9 & t_{13} & t_{17} & t_{21}\\
t_1+1 & t_6 & t_{10} & t_{14} & t_{18} & t_{22}\\
t_2 & t_5+1 & t_{10}+1 & t_{15} & t_{19} & t_{22}\\
t_2+1 & t_6+1 & t_9+1 & t_{16} & t_{20} & t_{21}\\
t_3 & t_7 & t_{11} & t_{13}+1 & t_{20}+1 & t_{22}+1\\
t_3+1 & t_8 & t_{12} & t_{16}+1 & t_{17}+1 & t_{22}+1\\
t_4 & t_7+1 & t_{12}+1 & t_{14}+1 & t_{19}+1 & t_{21}+1\\
t_4+1 & t_8+1 & t_{11}+1 & t_{15}+1 & t_{18}+1 & t_{21}+1
\end{pmatrix}
\begin{pmatrix}
t_1 & t_5 & t_9 & t_{13} & t_{17} & t_{21}\\
t_1+1 & t_6 & t_{10} & t_{14} & t_{18} & t_{21}\\
t_2 & t_5+1 & t_{11} & t_{15} & t_{18}+1 & t_{22}\\
t_2+1 & t_6+1 & t_9+1 & t_{16} & t_{19} & t_{22}\\
t_3 & t_7 & t_{10}+1 & t_{13}+1 & t_{20} & t_{22}+1\\
t_3+1 & t_8 & t_{12} & t_{14}+1 & t_{17}+1 & t_{22}+1\\
t_4 & t_7+1 & t_{12}+1 & t_{15}+1 & t_{19}+1 & t_{21}+1\\
t_4+1 & t_8+1 & t_{11}+1 & t_{16}+1 & t_{20}+1 & t_{21}+1
\end{pmatrix}$$
$$\text{22 parameters}, |Aut| = 16 \qquad\qquad \text{22 parameters}, |Aut| = 16$$

$$\begin{pmatrix}
t_1 & t_5 & t_9 & t_{13} & t_{17} & t_{21}\\
t_1+1 & t_6 & t_{10} & t_{14} & t_{18} & t_{22}\\
t_2 & t_5+1 & t_{11} & t_{15} & t_{19} & t_{22}+1\\
t_2+1 & t_6+1 & t_9+1 & t_{16} & t_{20} & t_{21}\\
t_3 & t_7 & t_{10}+1 & t_{13}+1 & t_{20}+1 & t_{22}\\
t_3+1 & t_8 & t_{12} & t_{14}+1 & t_{19}+1 & t_{21}+1\\
t_4 & t_7+1 & t_{12}+1 & t_{15}+1 & t_{18}+1 & t_{21}+1\\
t_4+1 & t_8+1 & t_{11}+1 & t_{16}+1 & t_{17}+1 & t_{22}+1
\end{pmatrix}
\begin{pmatrix}
t_1 & t_5 & t_9 & t_{13} & t_{17} & t_{21}\\
t_1+1 & t_6 & t_{10} & t_{14} & t_{18} & t_{21}\\
t_2 & t_5+1 & t_{11} & t_{15} & t_{18}+1 & t_{22}\\
t_2+1 & t_6+1 & t_9+1 & t_{16} & t_{19} & t_{22}\\
t_3 & t_7 & t_{10}+1 & t_{13}+1 & t_{20} & t_{22}+1\\
t_3+1 & t_8 & t_{12} & t_{15}+1 & t_{19}+1 & t_{21}+1\\
t_4 & t_7+1 & t_{12}+1 & t_{14}+1 & t_{17}+1 & t_{22}+1\\
t_4+1 & t_8+1 & t_{11}+1 & t_{16}+1 & t_{20}+1 & t_{21}+1
\end{pmatrix}$$
$$\text{22 parameters}, |Aut| = 32 \qquad\qquad \text{22 parameters}, |Aut| = 8$$

$$\begin{pmatrix}
t_1 & t_5 & t_9 & t_{13} & t_{17} & t_{21}\\
t_1+1 & t_6 & t_{10} & t_{14} & t_{18} & t_{22}\\
t_2 & t_5+1 & t_{11} & t_{15} & t_{19} & t_{22}+1\\
t_2+1 & t_6+1 & t_9+1 & t_{16} & t_{20} & t_{21}\\
t_3 & t_7 & t_{10}+1 & t_{13}+1 & t_{20}+1 & t_{22}\\
t_3+1 & t_8 & t_{12} & t_{15}+1 & t_{18}+1 & t_{21}+1\\
t_4 & t_7+1 & t_{12}+1 & t_{14}+1 & t_{19}+1 & t_{21}+1\\
t_4+1 & t_8+1 & t_{11}+1 & t_{16}+1 & t_{17}+1 & t_{22}+1
\end{pmatrix}$$
$$\text{22 parameters}, |Aut| = 16$$

Fig. 12.7 The non-extensible 6-dimensional combinatorial cube packings with 8 cubes and at least 21 parameters

(ii) $8 = f^T_\infty(6) < f^T_{>0,\infty}(6)$.

Proof. (ii) follows immediately from (i). The enumeration problem in (i) is solved in the following way: instead of adding cube after cube like in the random cube packing process, we add coordinate after coordinate in all possible ways and reduce by isomorphism. The computation returns the listed combinatorial torus cube packings. Given a combinatorial cube packing $\mathcal{CP}$ in order to prove that $p(\mathcal{CP}, \infty) = 0$, we consider all $(8!)$ possible paths p and see that for all of them $p(\mathcal{CP}, p, \infty) = 0$.   □

Proposition 12.6. *If $n = 3, 5, 7, 9$, then there exists a combinatorial torus cube tiling obtained with strictly positive probability and $\frac{n(n+1)}{2}$ parameters.*

Proof. If n is odd consider the matrix $H_n = m_{i,j}$ with all elements satisfying $m_{i+k,i} = m_{i-k,i} + 1$ for $1 \leq k \leq \frac{n-1}{2}$, the addition being modulo n. The matrix for $n = 5$ is

$$H_5 = \begin{pmatrix} t_1 & t_7 + 1 & t_{13} + 1 & t_{14} & t_{10} \\ t_6 & t_2 & t_8 + 1 & t_{14} + 1 & t_{15} \\ t_{11} & t_7 & t_3 & t_9 + 1 & t_{15} + 1 \\ t_{11} + 1 & t_{12} & t_8 & t_4 & t_{10} + 1 \\ t_6 + 1 & t_{12} + 1 & t_{13} & t_9 & t_5 \end{pmatrix}.$$

Then form the combinatorial cube packing with the cubes $(z^i + [0,1]^n)_{1 \leq i \leq n}$ and z^i being the i-th row of H_n. It is easy to see that the number of parameters of cubes, which we can add after z^i is $n - i$. So, those first n cubes are attained with the minimal number $\frac{n(n+1)}{2}$ of parameters and with strictly positive probability. If a cube $z + [0,1]^n$ is non-overlapping with $z^i + [0,1]^n$ for $i \leq n$ then there exists $\sigma(i) \in \{1, \dots, n\}$ such that $z_{\sigma(i)} = z^i_{\sigma(i)} + 1$. If $i \neq i'$ then $\sigma(i) \neq \sigma(i')$, which proves that $\sigma \in \text{Sym}(n)$. We also add the n cubes corresponding to the matrix $H_n + Id_n$. So, there are $n!$ possibilities for adding new cubes and we need to prove that we can select $2^n - 2n$ non-overlapping cubes amongst them.

The symmetry group of the n cubes $(z^i + [0,1]^n)_{1 \leq i \leq n}$ is the dihedral group D_{2n} with $2n$ elements. It acts on $\text{Sym}(n)$ by conjugation and so we simply need to list the relevant set of inequivalent permutations in order to describe the corresponding cube packings. See Table 12.5 for the found permutation for $n = 3, 5, 7, 9$.   □

The cube packing of above theorem was obtained for $n = 5$ by random method, i.e., adding cube whenever possible by choosing at random. Then the packings for $n = 7$ and 9 were built using the matrix H_n and consideration of all possibilities invariant under the dihedral group D_{2n} by computer. But for $n = 11$ this method does not work. It would be interesting to know in which dimensions n combinatorial torus cube tilings with $\frac{n(n+1)}{2}$ parameters do exist.

Table 12.5   List of permutation describing combinatorial torus cube tilings with $\frac{n(n+1)}{2}$ parameters in dimension 3, 5, 7, 9

$n = 3$	$(1,2,3)$		
$n = 5$	$(1,2,3,4,5)$	$(1,2)(3,5,4)$	$(1,4,5,3,2)$
$n = 7$	$(1,2,3,4,5,6,7)$	$(1,7)(2,5,4,3,6)$	$(1,6,2,5,4,3,7)$
	$(1,7)(2,5,6)(3,4)$	$(1,6,2,3,7)(4,5)$	$(1,7)(2,3,4,5,6)$
	$(1,3,7)(2,6)(4,5)$	$(1,3,7)(2,5,4,6)$	$(1,5,4,3,2,6,7)$
$n = 9$	$(1,2,3,4,5,6,7,8,9)$	$(1,6,7,4,3,5,9)(2,8)$	$(1,5,6,7,4,3,9)(2,8)$
	$(1,5,9)(2,8)(3,6,7,4)$	$(1,9)(2,5,4,3,6,7,8)$	$(1,6,7,8,2,5,4,3,9)$
	$(1,5,4,3,9)(2,8)(6,7)$	$(1,9)(2,5,6,7,8)(3,4)$	$(1,9)(2,3,6,5,4,7,8)$
	$(1,6,5,4,7,8,2,3,9)$	$(1,9)(2,3,6,7,8)(4,5)$	$(1,6,7,8,2,3,9)(4,5)$
	$(1,9)(2,3,4,7,6,5,8)$	$(1,9)(2,3,4,5,6,7,8)$	$(1,9)(2,3,6)(4,7,8,5)$
	$(1,6,2,3,9)(4,7,8,5)$	$(1,5,4,7,8,6,2,3,9)$	$(1,8,3,7,6,4,9)(2,5)$
	$(1,7,6,4,9)(2,5)(3,8)$	$(1,7,6,9)(2,8,3,4,5)$	$(1,4,5,2,8,3,7,6,9)$
	$(1,4,9)(2,5)(3,7,6,8)$	$(1,4,8,3,7,6,9)(2,5)$	$(1,7,6,3,4,5,2,8,9)$
	$(1,4,5,2,8,9)(3,7,6)$	$(1,4,9)(2,5)(3,8,7,6)$	$(1,3,7,6,9)(2,8,4,5)$
	$(1,7,6,5,4,3,2,8,9)$	$(1,9)(2,5,8)(3,4)(6,7)$	

Appendix A

Combinatorial Enumeration

We consider here exhaustive combinatorial problems, i.e. all elements of a finite class of combinatorial objects. Example of problem considered are

- List all 3-regular plane graphs with faces of size 5 and 9 and all 9-gonal faces in pairs (see [Deza and Dutour Sikirić (2008)] for the context).
- List all independent sets of 600-cell (see [Dutour Sikirić and Myrvold (2008)] for the context).
- List all triangulations of the sphere on n vertices (see [Brinkmann and McKay (2007)] for the method).
- List all cube tiling of $[0, 4]^n$ by integral translates of the cube $[0, 2]^n$.

The main feature of the proposed problems is that we do not have any intelligent way of doing it. Also, we do not want to know only the numbers of objects, we want to have those objects so as to work with them. One of the possible uses of this is for building infinite series from found examples. More generally science progress by examples and classifications and for combinatorial research, having new examples to study is absolutely essential.

In the best scenario, the speed of computers multiply by 2 every year. In most combinatorial problems, the number of solutions grows much more than exponentially in the size of the problem. One typical example is the listing of all graphs with n vertices: The number of labeled graphs with n vertices is $2^{\frac{n(n-1)}{2}}$. The symmetric group $Sym(n)$ acts on those labeled graphs. So, the number of unlabeled graphs, i.e. graphs up to isomorphism, is around

$$2^{\frac{n(n-1)}{2}} \frac{1}{n!} \simeq \sqrt{2}^{n^2}.$$

As a consequence the benefit brought by computer should diminish as time goes on.

There are a number of classic technique in combinatorial enumeration that we are not considering. Formal power series and generating functions are considered in Chapter 2. There are also many such techniques in Statistical Physics. Another method we do not describe is the Polya/Redfield method: it consists in enumerating the possible group of symmetry and for every group to list the possible structures. There exist also in some problems special techniques that are faster than pure combinatorial enumeration but which are still computationally intensive, one example is polyhedral computations in [Schürmann (2009)]. The methods explained here are generally bad and are to be used when no other structure is present.

The applications of exhaustive enumeration besides obtaining classification result is to get new remarkable object, when no other method is available and to find infinite series from the first instances. Variants of the method are enumerating with respect to a fixed symmetry group, and the branch and bound methods of combinatorial optimization using symmetries [Margot (2003)].

First we present the automorphism and isomorphism problems, then sequential enumeration until finally we expose orderly enumeration techniques and the homomorphism principle.

A.1　The isomorphism and automorphism problems

We first begin by the graph automorphism problem.

Suppose that we have a graph G on n vertices $\{1, \ldots, n\}$, we want to compute its automorphism group $\mathrm{Aut}(G)$ that is the group of permutations $\sigma \in Sym(n)$ such that

$$\{\sigma(i), \sigma(j)\} \in E(G) \text{ if and only if } \{i, j\} \in E(G)$$

with $E(G)$ the edge set of G. Suppose that G_1 and G_2 are two graphs on n vertices $\{1, \ldots, n\}$, we want to test if G_1 and G_2 are isomorphic, i.e. if there is $g \in Sym(n)$ such that

$$\{g(i), g(j)\} \in E(G_1) \text{ if and only if } \{i, j\} \in E(G_2).$$

It is generally believed that those problems do not admit solution in a time bounded by a polynomial in n, see [Köbler, Schöning and Torán (1993)] for an extensive discussion of the graph isomorphism problem.

The program nauty by Brendan McKay solves the graph isomorphism and the automorphism problems [McKay (1981); MacKay (2008)]. nauty is extremely efficient in doing those computations and can deal with vertex colored graph. Yet the method used is really at bottom to iterate over all possible isomorphism, what is done subtly is the partitioning of the vertex set and the branching techniques. In practice **nauty** has no problem at all for graph with several hundred vertices. So, the above theoretical limitation is not a real one in practice.

At this point, the reader might be surprised that we focus on graphs when we promised to deal with combinatorial structures like subset of vertex-set of a graph, set system, edge weighted graph, plane graph, partially ordered set, etc. The idea is that such structures can be reduced to graphs and the isomorphism and automorphism problems reduced to a graph problem. More precisely, if M is a "combinatorial structure", then we have to define a graph $G(M)$, such that:

(i) If M_1 and M_2 are two "combinatorial structure", then M_1 and M_2 are isomorphic if and only if $G(M_1)$ and $G(M_2)$ are isomorphic.

(ii) If M is a "combinatorial structure", then $\mathrm{Aut}(M)$ is isomorphic to $\mathrm{Aut}(G(M))$.

Of course the graph $G(M)$ should have as few vertices as possible and the art is in choosing the right graph reduction for a given problem. Let us now consider a few cases:

Subset of vertex-set of a graph. Suppose that we have a graph G, two subsets S_1, S_2 of G, we want to know if there is an automorphism ϕ of G such that $\phi(S_1) = S_2$.

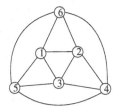

$$S_1 = \{1, 2, 4\}$$
$$S_2 = \{3, 5, 6\}$$

The method is to define two graphs associated to it:

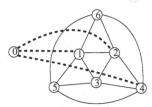 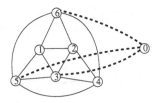

We give a different color to the vertex 0.

Set systems. Suppose we have some subsets $S_1, \ldots, S_r$ of $\{1, \ldots, n\}$. We want to find the permutations of $\{1, \ldots, n\}$, which permutes the S_i.

We define a graph with $n + r$ vertices j and S_i with j adjacent to S_i if and only if $j \in S_i$. Example $\mathcal{S} = \{\{1, 2, 3\}, \{1, 5, 6\}, \{3, 4, 5\}, \{2, 4, 6\}\}$:

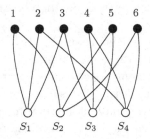

We give different colors to the vertices S_i and the vertices j.

Edge colored graphs. G is a graph with vertex-set $(v_i)_{1 \leq i \leq N}$, edges are colored with k colors $C_1, \ldots, C_k$:

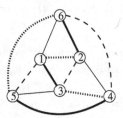

We want to find automorphisms preserving the graph and the edge colors. We form the graph with vertex-set (v_i, C_j) and

(1) edges between (v_i, C_j) and $(v_i, C_{j'})$,
(2) edges between (v_i, C_j) and $(v_{i'}, C_j)$ if there is an edge between v_i and $v_{i'}$ of color C_j.

We get a graph with kN vertices.

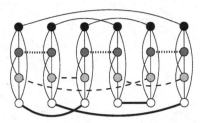

Actually, one can do better, if the binary expression of j is $b_1 \ldots b_r$ with $b_i = 0$ or 1 then we form the graph with vertex-set (v_i, l), $1 \leq l \leq r$ and

• edges between (v_i, l) and (v_i, l'),

- edges between (v_i, l) and $(v_{i'}, l)$ if the binary number b_l of the expression of C_j is 1.

This makes a graph with $\lceil \log_2(k) \rceil N$ vertices. Another possibility, which is actually less competitive, is to consider the line graph $L(\mathsf{K}_n)$ of the complete graph K_n. It is the graph defined on the edge set of K_n with two edges adjacent if they share an edge. By doing this we get from an edge colored graph a vertex colored one. The problem is that the number of vertices is $\frac{n(n-1)}{2}$ which is high. Another small problem is that $|\operatorname{Aut}(\mathsf{K}_4)| = 24$ and $|\operatorname{Aut}(L(\mathsf{K}_4))| = 48$, that is the line graph has some symmetries that were not present in the original graph. This indicates that one should be careful when doing such transformations.

Plane graphs. A plane graph is a graph that can be drawn on the plane with no two edges intersecting each other. A graph is k-connected if after removing any $k - 1$ vertices, it remains connected. The *skeleton* of a plane graph is the graph obtained by forgetting about the embedding, i.e. considering only the vertices and the edges. If G is a 3-connected plane graph with no multiple edges or loops then the skeleton determine the embedding and so we can apply nauty directly. If G has multiple edge and/or is not 3-connected then we consider the graph formed by its vertices, edges and faces with adjacency given by incidence. See below an example of a non-3-connected plane graph and the associated graph used for automorphism computations.

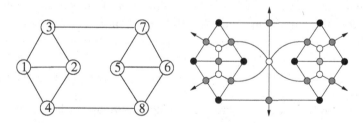

We should assign different colors to the vertices associated to vertices, edges or faces. Otherwise, we could get in the graph $G(M)$ some symmetries arising from self-duality. This construction extends to partially ordered sets, face lattices, etc.

nauty has yet another wonderful feature: it can compute a canonical form of a given graph. One possible canonical form of a graph is obtained by taking the lexicographic minimum of all possible adjacency matrices of a given graph. This canonical form cannot be computed efficiently and

nauty uses another one. Suppose that one has N different graphs from which we want to select the non-isomorphic ones. If one do isomorphism tests with **nauty** then at worst we have $\frac{N(N-1)}{2}$ tests. If one computes canonical forms, then we have N calls to **nauty** and then string equality tests. This feature is the key to many computer enumeration goals.

In conclusion we shall say that computing the automorphism group of a combinatorial structure is in general easy. The only difficulty is that one has to be careful in defining the graph $G(M)$. In many cases, most of the computational time is taken by the slow program writing the graph to a file.

A.2   Sequential exhaustive enumeration

The sequential exhaustive enumeration idea is to decompose the sought combinatorial structure into small blocks and to reconstruct it step by step. Suppose for example that one wants to enumerate the connected graphs on 10 vertices which are of degree 3 and with no multiple edges and loops. The starting point is to take a vertex v and to add to it three vertices v_1, v_2, v_3. Then, for the vertex v_1, we have to add two edges, we can take the vertices v_2, v_3 or add new vertices. The process continues and at the N-th stage of the enumeration, we have, say, $m(N)$ possible configurations.

The great problem of this enumeration scheme is that the number of cases to consider grows at each step. If the number of objects to enumerate is large, then this is unavoidable. But sometimes, we can have very large computations yielding a few or no objects. This is what makes combinatorial enumeration difficult. The enumeration scheme, i.e. the method by which we add objects has to be carefully chosen so as to finish in a reasonable time.

Besides choosing the right incremental scheme, two methods can help to speed up the computation: one is to use symmetries, i.e. to consider only objects which are resolved by isomorphism at the N-th step, another is to remove from consideration intermediate objects that we know will not yield combinatorial structures. For example a 3-regular graph with n vertices has $\frac{3}{2}n$ edges. If an intermediate structure has more than this number, then it cannot be completed.

So, the basic enumeration algorithm is to start from an empty structure and add component to it step by step, while checking that the obtained structures are consistent and have a possibility of generating full structures.

Note that sometimes the isomorphism tests take too much time with respect to other operations with negligible benefit. Thus it is sometimes preferable to do isomorphism tests only in the first few stage of the enumeration.

The time of run is unpredictable as well as the number of intermediate structures. Furthermore the symmetry of the obtained objects cannot be used if they are not known *a priori*. The method is essentially a computerized case by case analysis as it is done traditionally in many situations.

The above method has the disadvantage of storing all considered structures in memory, which is computationally intensive. Some methods exist that use symmetry, are based on tree search and are not memory intensive. This is called "Augmentation scheme", "Orderly generation" or "Isomorph free generation" and we refer to [Colbourn and Read (1979); Brinkmann and McKay (2007); Read (1978); Kaski and Ostergard (2006); Faradzev (1978); McKay (1998)] for more details. The clique case allow for specialized algorithms, see [Myrvold and Fowler (2007); Ostergard (2005); Dutour Sikirić and Myrvold (2008)]. Spectacular results have been obtained by using Orderly generation, see for example [Brinkmann and McKay (2002, 2007); Dinitz, Garnick and McKay (1994)].

A.3 The homomorphism principle

Suppose that one wants to generate 4-regular plane graphs with faces of size 2, 4, 6 such that every vertex is contained in exactly one face of size 2.

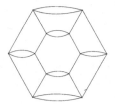 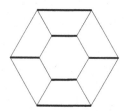

If one collapses the 2-gons to edges, one obtains a $(\{4, 6\}, 3)$-sphere. The 2-gons correspond to a perfect matching in it. The method is then

- List all $(\{4, 6\}, 3)$-graphs.
- For every $(\{4, 6\}, 3)$-graph, list its perfect matching.

By doing so, we reduce considerably the difficulties.

The method is thus in general to consider a substructure to the given problem and to enumerate it and then progressively enrich it. A more detailed discussion of the practical algebraic computation one has to consider when doing combinatorial enumeration is available from [Brinkmann (2000); Kerber (1999)].

Bibliography

Abramowitz, M. and Stegun, I. A.. (1965). *Handbook of Mathematical Functions*, Dover, New York.

Akeda, Y. and Hori, M. (1975). Numerical test of Palásti's conjecture on two-dimensional random packing density, *Nature* **254**, pp. 318–319.

Akeda, Y. and Hori M. (1976). On random sequential packing in two and three dimensions, *Biometrika* **63**, pp. 361–366.

Alspach, B. (2008). The wonderful Walecki construction, http://www.math.mtu.edu/~kreher/ABOUTME/syllabus/Walecki.ps

Andreatta, M. (2006). On group-theoretical methods applied to music: some compositional and implementational aspects, to appear in *Perspectives of Mathematical and Computer-Aided Music Theory*, University of Osnabrück, edited by E. Lluis-Puebla, G. Mazzola and T. Noll.

Ash, R. (1965). *Information Theory*, Wiley, New York.

Aste, T. and Weaire, D. (2008). *The Pursuit of Perfect Packing (second edition)*, Taylor & Francis.

Avez, A. (1986). *Differential Calculus*, Wiley, New York.

Ballinger, B., Blekherman, G., Cohn, H., Giansiracusa, N., Kelly, E. and Schürmann, A. (2009). Experimental study of energy-minimizing point configurations on spheres, *Experiment. Math.* **18**, pp. 257–283.

Balser, W. (1999). *Formal Power Series and Linear Systems of Meromorphic Differentials*, Springer-Verlag, New York.

Bankövi, G. (1962). On gaps generated by a random space filling procedure, *Publ. Math. Inst. Hungar. Acad. Sci.* **7**, pp. 395–407.

Bannai, E. and Sloane, N. J. A. (1981). Uniqueness of certain sperical codes, *Can. J. Math.* **33**, pp. 437–449.

Barbour, A. D., Holst, L. and Janson, S. (1992). *Poisson Approximation*, Oxford University Press, Oxford.

Baryshnikov, Y. and Yukich, J.E. (2002). Gaussian field and random packing, *J. Stat. Phys.* **111**, pp. 443–463.

Bernal, J. D. (1959). A geometrical approach to the structure of liquids, *Nature* **17**, pp. 141–147.

Blaisdell, B. E. and Solomon, H. (1970). On a random sequential packing in the plane and a conjecture of Palásti, *J. Appl. Prob.* **7**, pp. 667–698.

Blaisdell, B. E. and Solomon, H. (1982). Random sequential packing in Euclidean spaces of dimensions three and four and a conjecture of Palásti, *J. Appl. Prob.* **19**, pp. 382–390.

Boissonnat, J. D., Sharir, M., Tagansky, B. and Yvinec, M. (1998). Voronoi diagrams in higher dimensions under certain polyhedral distance functions, *Discrete Comput. Geom.* **19**, pp. 485–519.

Bölcskei, A. (2000). Classification of unilateral and equitransitive tilings by squares of three sizes, *Beiträge zur Alg. und Geom.* **41**, pp. 267–277.

Bölcskei, A. (2001). Filling space with cubes of two sizes, *Publ. Math. Debrecen* **59**, pp. 317–326.

Böröczky, K. (2004). *Finite Packing and Covering*, Cambridge University Press, Cambridge.

Brinkmann, G. (2000). Isomorphism rejection in structure generation programs, *Discrete mathematical chemistry, DIMACS Ser. Discrete Math. Theoret. Comput. Sci.* (Amer. Math. Soc.) **51**, pp. 25–38.

Brinkmann, G. and McKay, B. D. (2007). Fast generation of planar graphs, *MATCH Commun. Math. Comput. Chem.* **58**, pp. 323–357.

Brinkmann, G. and McKay, B. D. (2002). Posets on up to 16 points, *Order* **19**, pp. 147–179.

Cesari, L. (1963). *Asymptotic Behavior and Stability Problems in Ordinary Differential Equations*, Springer-Verlag, Berlin.

Chew, L. P., Kedem, K., Sharir, M., Tagansky, B. and Welzl, E. (1998). Voronoi diagrams of lines in three dimensions under polyhedral convex distance functions, *J. Algorithms* **29**, pp. 238–255.

Coddington, E. A. and Levinson, N. (1955). *Theory of Ordinary Differential Equations*, McGraw-Hill.

Coffman, E. G., Flatto, L., Jelenković, P. and Poonen, B. (1998). Packing Random Intervals On-Line. Average case analysis of algorithms, *Algorithmica* **22**, pp. 448–476.

Coffman, E. G., Jelenković, P. and Poonen, B. (1999). Reservation probabilities, *Adv. Performance Analysis* **2**, pp. 129–158.

Coffman, E. G., Mallows, C. L. and Poonen, B. (19994). Parking arcs on the circle with applications to one-dimensional communication networks, *Ann. Appl. Probab.* **4**, pp. 1098–1111.

Coffman, E. G., Poonen, B. and Winkler, P. M. (1995). Packing random intervals, *Probab. Theory Related Fields* **102**, pp. 105–121.

Cohen, G., Honkala, I., Litsyn, S. and Lobstein, A. (1997). *Covering Codes*, North Holland Mathematical Library.

Cohen, E. R. and Reiss, H. (1963). Kinetics of Reactant Isolation. I. One-dimensional problems, *J. Chem. Phys.* **38**, pp. 680–691.

Colbourn, C. J. and Read, R. C. (1979). Orderly algorithms for graph generation, *Internat. J. Comput. Math.* **7**, pp. 167–172.

Conway, J. H., Delgado Friedrichs, O., Huson, D. H. and Thurston, W. P. (2001). On three-dimensional space groups, *Beiträge zur Alg. und Geom.* **42**, pp. 475–507.

Conway, J. H. and Pless, V. (1980). On the enumeration of self-dual codes, *J. Comb. Th. A* **28**, pp. 26–53.

Conway, J. H. and Sloane, N. J. A. (1999). *Sphere Packings, Lattices and Groups (third edition)*, Springer-Verlag, New York.

Corrádi, K. and Szabó, S. (1990). A combinatorial approach for Keller's conjecture, *Period. Math. Hungar.* **21**, pp. 95–100.

Deligne, P. (1970). *Équation Différentielle à Points Singuliers Régulier*, Springer-Verlag, Paris.

Delsarte, P. and Goethals, J. M. (1975). Unrestricted codes with the Golay parameters are unique. *Disc. Math.* **12**, pp. 211–224.

Devroye, L. (1986), A note on the height of binary search trees, *J. Assoc. Comput. Mach.* **33**, pp. 489–498.

Deza, M. and Dutour Sikirić, M. (2008). *Geometry of Chemical Graphs*, Cambridge University Press, Cambridge.

Diaconis, P. (2008), The Markov chain Monte Carlo revolution, *Bull. Amer. Math. Soc.* **46**, pp. 179–205.

Dickson, L. E. and Safford, F. H. (1906). Solution to problem 8 (group theory), *Amer. Math. Month.* **13**, pp. 150–151.

Dinitz, J. H., Garnick, D. K. and McKay, B. D. (1994). There are 526915620 nonisomorphic one-factorizations of K_{12}, *J. Combin. Des.* **2**, pp. 273–285.

Dolbilin, N., Itoh, Y. and Poyarkov, A. (2005). On random tilings and packings of space by cubes, *The Proceedings of COE workshop on sphere packings*, Kyushu University, Fukuoka, pp. 70–79.

Dolby, J. L. and Solomon, H. (1975). Information density phenomena and random packing, *J. Appl. Prob.* **12**, pp. 364–370.

Donev, A., Torquato, S., Stillinger, F.H. and Connelly, R. (2004). A linear programming algorithm to test for jamming in hard-sphere packings. *J. Comput. Phys.* **197**, pp. 139–166.

Drmota, M. (2009). *Random Trees, An Interplay between Combinatorics and Probability.* Springer-Verlag, Berlin.

Dutour Sikirić, M. (2008). Programs for continuous cube packings, http://www.liga.ens.fr/~dutour/Programs.html

Dutour Sikirić, M. and Itoh, Y. (2010). Combinatorial cube packings in the cube and the torus, *Eur. J. Combin.* **31**, pp. 517–534.

Dutour Sikirić, M., Itoh, Y. and Poyarkov, A. (2007). Cube packings, second moment and holes, *Eur. J. Combin.* **28**, pp. 715–725.

Dutour Sikirić, M. and Myrvold, W. (2008). The special cuts of 600-cell, *Beiträge zur Alg. und Geom.* **49**, pp. 269–275

Dutour Sikirić, M., Schürmann, A. and Vallentin, F. (2009). Complexity and algorithms for computing Voronoi cells of lattices, *Math. Comp.* **78**, pp. 1713–1731.

Dvoretzky, A. (1956). On covering a circle by randomly placed arcs, *Proc. Nat. Acad. Sci. U.S.A.* **42**, pp. 199–203.

Dvoretzky, A. and Robbins, H. (1964). On the "parking problem", *Publ. Math. Inst. Hungar. Acad. Sci.* **9**, pp. 209–225.

Eick, B. and Souvignier, B. (2006). Algorithms for crystallographic groups, *Int.*

J. Quant. Chem. **106**, pp. 316–343.

Evans, J. W. (1993). Random and cooperative sequential adsorption, *Rev. Mod. Phys.* **65**, pp. 1281–1329.

Faradzev, I. A. (1978). Generation of nonisomorphic graphs with a given distribution of the degrees of vertices (in Russian), *Algorithmic Studies in Combinatorics "Nauka", Moscow* **185**, pp. 11–19.

Fejes Tóth, G., Kuperberg, G. and Kuperberg, W. (1998). Highly saturated packings and reduced coverings, *Monatsh. Math.* **125**, pp. 127–145.

Fejes Tóth, G. and Kuperberg, W. (1993). Packings and coverings with convex sets, *Handbook of Convex Geometry, Vol A, B*, pp. 799–860, North Holland.

Feller, W. (1968). *An Introduction to Probability Theory and its Applications I.*, Wiley, New York.

Feller, W. (1971). *An Introduction to Probability Theory and its Applications II.*, Wiley, New York.

Flajolet, P. and Odlyzko, A. (1982). The average height of binary trees and other simple trees, *J. Comp. Syst. Sci.* **25**, pp. 171–213.

Flajolet, P. and Sedgewick, R. (2009). *Analytic Combinatorics*, Cambridge University Press, Cambridge.

Flatto, L. and Konheim, A. G. (1962). The random division of an interval and the random covering of a circle, *SIAM Rev.* **4**, pp. 211–222.

Flory, P. J. (1939). Intramolecular Reaction between Neighboring Substituents of Vinyl Polymers, *J. Am. Chem. Soc.* **61**, pp. 1518–1521.

The GAP Group, GAP — Groups, Algorithms, and Programming, Version 4.3; 2002. http://www.gap-system.org.

Gelling, E. N. and Odeh, R. E. (1974). On 1-factorizations of the complete graph and the relationship to round-robin schedule, *Congr. Numerant.* **9**, pp. 213–221.

Golay, M. J. E. (1949). Notes on digital coding, *Proc. I.R.W.(I.E.E.E.)* **37**, pp. 657.

Graham, R., Knuth, D. E. and Patashnik, O. (1994). *Concrete Mathematics. A Foundation for Computer Science*, Addison Wesley.

Grünbaum, B. and Shepard, G. C. (1987). *Tilings and Patterns*, W. H. Freeman pp. 72–81.

Hahn, T., ed. (2002). *International Tables for Crystallography, Volume A: Space Group Symmetry* (5th ed.), Springer-Verlag, Berlin.

Hales, T. C. (2000). Cannonballs and honeycombs, *Notices. Amer. Math. Soc.* **47**, pp. 440–449.

Hales, T. C. (2005). A proof of the Kepler conjecture, *Ann. Math.* **162**, pp. 1065–1185.

Hamming, R. W. (1947). *Self-Correcting Codes-Case 20878*, Memorandum 1130-RWHMFW, Bell Telephone Laboratories.

Harary, F., Robinson, R. W. and Wormald, N. C. (1978). Isomorphic factorisations. I. Complete graphs, *Trans. Amer. Math. Soc.* **242**, pp. 243–260.

Hardy, G. H. and Littlewood, J. E. (1914). Tauberian theorems concerning power series and Dirichlet's series whose coefficients are positive, *Proc London Math. Soc. (2)* **13**, pp. 174–191.

Hardy, G. H. and Littlewood, J. E. (1930). Notes on the theory of series (XI): on Tauberian theorems, *Proc. London Math. Soc. (2)* **30**, pp. 23–37.

Hattori, T. and Ochiai, H. (2006). Scaling Limit of successive approximations for $w' = -w^2$, *Funkcial. Ekvac.* **49**, pp. 291–319.

Heden, O. (2008). A survey of perfect codes, *Adv. Math. Commun.* **2**, pp. 223–247.

Hemmer, P. C. (1989). The Random Parking Problem, *J. Stat. Phys.* **57**, pp. 865–869.

Higuti, I. (1960). A statistical study of random packing of unequal spheres, *Ann. Inst. Statist. Math.* **12**, pp. 257–271.

Ince, E. L. (1944). *Ordinary Differential Equations*, Dover publications.

Itoh, Y. (1978). Random packing model for nomination of candidates and its application to elections of the House of Representatives in Japan, *Proc. Internat. Conf. Cybernetics and Society* **1**, pp. 432–435.

Itoh, Y. (1980). On the minimum of gaps generated by one-dimensional random packing, *J. Appl. Prob.* **17**, pp. 134–144.

Itoh, Y. (1985). Note on a restricted random cutting of a stick (in Japanese), *Proc. Inst. Statist. Math.* **33**, pp. 97–99.

Itoh, Y. (1985). Random packing by Hamming distance (In Japanese), *Proc. Inst. Statist. Math.* **33**, pp. 156–157.

Itoh, Y. (1986). Golay code and random packing, *Ann. Inst. Statist. Math.* **38**, pp. 583–588.

Itoh, Y. and Hasegawa, M. (1980). Why twenty amino acids (in Japanese), *Seibutsubutsuri* **21**, pp. 21–22.

Itoh, Y. and Jimbo, M. (1987). A stochastic construction of Golay code, *Designs and finite geometries (Kyoto, 1986) Surikaisekikenkyusho Kokyuroku* **607**, pp. 76–82.

Itoh, Y. and Mahmoud, H. M. (2003). One sided variations on interval trees, *J. Appl. Prob.* **40**, pp. 654–670.

Itoh, Y. and Shepp, L. (1999). Parking Cars with Spin but no Length, *J. Stat. Phys.* **97**, pp. 209–231.

Itoh, Y. and Solomon, H. (1986). Random sequential coding by Hamming distance, *J. Appl. Prob.* **23**, pp. 688–695.

Itoh, Y. and Ueda, S. (1978). Note on random packing models for an analysis of elections (in Japanese), *Proc. Inst. Statist. Math.* **25**, pp. 23–27.

Itoh, Y. and Ueda, S. (1979). A random packing model for elections, *Ann. Inst. Statist. Math.* **31**, pp. 157–167.

Itoh, Y. and Ueda, S. (1983). On packing density by a discrete random sequential packing of cubes in a space of n dimension (in Japanese), *Proc. Inst. Statist. Math.* **31**, pp. 65–69.

Jaeger, H. M. and Nagel, S. R. (1992). Physics of Granular States, *Science* **255**, pp. 1523–1531.

Kahane, J. P. (1985). *Some Random Series of Functions (second edition)*, Cambridge Studies in Advanced Mathematics, Cambridge.

Kakutani, S. (1975). A problem of equidistribution on the unit interval $[0, 1]$, *Lecture Notes in Mathematics* **541**, pp. 369–376.

Kaski, P. and Ostergard, P. (2006). *Classification Algorithms for Codes and Designs*, Springer-Verlag, Berlin.

Kaski, P. and Ostergard, P. (2009). There are 1, 132, 835, 421, 602, 062, 347 nonisomorphic one-factorizations of K_{14}, *J. Combin. Des.* **17**, pp. 147–159.

Keller, O. H. (1930). Über die lückenlose Einfüllung des Raumes mit Würfeln, *J. Reine Angew. Math.* **163**, pp. 231–248.

Kerber, A. (1999). *Applied Finite Group Actions (second edition)*, Springer-Verlag, Berlin.

Knuth, D. E. (1973). *The Art of Computer Programming. Volume 3. Sorting and Searching*, Addison-Wesley.

Knuth, D. E. (1975). *The Art of Computer Programming. Volume 1: Fundamental Algorithms (second edition)*, Addison-Wesley.

Köbler, J., Schöning, U. and Torán, J. (1993). *The Graph Isomorphism Problem: its Structural Complexity*, Birkhäuser.

Komaki, F. and Itoh, Y. (1992). A unified model for Kakutani's interval splitting and Rényi's random packing, *Adv. Appl. Prob.* **24**, pp. 502–505.

Konheim, A. G. and Flatto, L. (1962). *SIAM Rev.* **4**, pp. 257–258.

Korevaar, J. (2004). *Tauberian Theory - A Century of Developments*, Springer-Verlag, Berlin.

Krapivsky, P. L. (1992). Kinetics of random sequential parking on a line, *J. Stat. Phys.* **69**, pp. 135–150.

Lagarias, J. C., Reeds, J. A. and Wang, Y. (2000). Orthonormal bases of exponentials for the n-cube, *Duke Math. J.* **103**, pp. 25–36.

Lagarias, J. C. and Shor, P. W. (1992). Keller's Cube tiling conjecture is false in High dimensions, *Bull. Amer. Math. Soc.* **27**, pp. 279–283.

Lagarias, L. C. and Shor, P. W. (1994). Cube-tilings of $\mathbb{R}^n$ and nonlinear codes, *Disc. Comput. Geom.* **11**, pp. 359–391.

Lagrange, J. L. (1773). Recherches d'arithmétiques, *Nouveaux Mémoires de l'Académie royale des sciences et Belles-Lettres de Berlin*, pp. 265–312.

Leech, J. (1964). Some sphere packings in Higher space, *Canad. J. Math.* **16**, pp. 657–682.

Lines, M. E. (1979). Hard-sphere random-packing model for an ionic glass: Yttrium iron garnet, *Phys. Rev. B* **20**, pp. 3729–3738.

Liu, J. S. (2001). *Monte Carlo Strategies in Scientific Computing*, Springer-Verlag, Berlin.

Lootgieter, J. C. (1977). Sur la répartition des suites de Kakutani. I, *Ann. Inst. H. Poincaré Sect. B (N.S.)* **13**, pp. 385–410.

Lynch, W. (1965). More combinatorial problems on certain trees, *The Computer J.* **7**, pp. 299–302.

Mackenzie, J. K. (1962). Sequential filling of a line by intervals placed at random and its application to linear adsorption, *J. Chem. Phys.* **37**, pp. 723–728.

Mackey, J. (2002). A cube tiling of dimension eight with no facesharing, *Disc. Comput. Geom.* **28**, pp. 275–279.

MacWilliams, E. J. and Sloane, N. J. A. (1977). *The Theory of Error-Correcting Codes I-II*, North-Holland.

Magnus, W., Oberhettinger, F. and Soni, R. P. (1966). *Formulas and Theorems for the Special Functions of Mathematical Physics*, Springer-Verlag, New York.

Mahmoud, H. (1992). *Evolution of Random Search Trees*, Wiley, New York.

Mahmoud, H. and Pittel, B. (1984). On the most probable shape of a binary search tree grown from a random permutation, *SIAM J. Alg. Disc. Meth.* **5**, pp. 69–81.

Mallows, C. L., Pless, V. and Sloane, N. J. A. (1976). Self-dual codes over $GF(3)$, *SIAM J. Appl. Math.* **31**, pp. 649–666.

Mandelbrot, B. (1972). On Dvoretzky coverings for the circle, *Z. Wahrscheinlichkeitstheorie und Verw. Gebiete* **22**, pp. 158–160.

Margot, F. (2003). Exploiting symmetries in symmetric ILP, *Mathematical programming, Ser. B* **98**, pp. 3–21.

Martinet, J. (2003). *Perfect Lattices in Euclidean Spaces*. Springer-Verlag, Berlin.

Martini, H., Makai, E. and Soltan, V. (1998). Unilateral tilings of the plane with squares of three sizes, *Beiträge zur Alg. und Geom.* **39**, pp. 481–495.

Matheson, A. J. (1974). Computation of a random packing of hard spheres, *J. Phys. C: Solid State Phys.* **7**, pp. 2569–2576.

McKay, B. D., Practical graph isomorphism, *Congr. Numerant.* **30**, pp. 45–87.

McKay, B. D. (1998). Isomorph-free exhaustive generation, *J. Algorithms* **26**, pp. 306–324.

McKay, B. D. (2008). *The nauty program*, http://cs.anu.edu.au/people/bdm/nauty/.

Morrison, J. A. (1987). The minimum of gaps generated by random packing of unit intervals into a large interval, *SIAM J. Appl. Math.* **47**, pp. 398–410.

Moser, L. and Wyman, M. (1958). Asymptotic development of the stirling numbers of the first kind, *J. London Math. Soc.* **33**, pp. 133–146.

Mounits, B., Etzion, T. and Litsyn, S. (2002). Improved Upper bounds on sizes of codes, *IEEE Trans. Inform. Th.* **48**, pp. 880–886.

Mounits, B., Etzion, T. and Litsyn, S. (2007). New upper bounds on codes via association schemes and linear programming, *Adv. Math. Comm.* **1**, pp. 173–195.

Myrvold, W. and Fowler, P. W. (2007). Fast enumeration of all independent sets of a graph, preprint.

Nebe, G. and Plesken, W. (1995). Finite rational matrix groups, *Mem. Amer. Math. Soc.* **116**.

Nebe, G. (1996). Finite subgroups of $GL_n(\mathbb{Q})$ for $25 \leq n \leq 31$, *Comm. Algebra* **24**, pp. 2341–2397.

Ney, P. E. (1962). A random space filling problem, *Ann. Math. Statist.* **33**, pp. 702–718.

O'Keeffe, M. and Hyde, B. G. (1996). *Crystal Structures I. Pattern and Symmetry*, Mineral. Soc. of America.

Opgenorth, J., Plesken, W. and Schulz, T. (1998). Crystallographic algorithms and tables, *Acta Cryst. A* **54**, pp. 517–531.

Ostergard, P. (2005). Constructing combinatorial objects via cliques, *Surveys in combinatorics*, 57–82, London Math. Soc. Lecture Note Ser., 327.

Palásti, I. (1960). On some random space filling problems, *Publ. Math. Inst. Hungar. Acad. Sci.* **5**, pp. 353–360.

Penrose, M. D. (2001). Random parking, sequential adsorption and the jamming limit, *Commun. Math. Phys.* **218**, pp. 153–176.

Penrose, M. D. and. Yukich, J. E. (2001). Central limit theorems for some graphs in computational geometry, *Ann. Appl. Prob.* **11**, pp. 1005–1041.

Penrose, M. D. and. Yukich, J. E. (2002). Limit theory for random sequential packing and deposition, *Ann. Appl. Prob.* **12**, pp. 272–301.

Penrose, M. D. and. Yukich, J. E. (2003). Weak laws of large numbers in geometric probability, *Ann. Appl. Prob* **13**, pp. 277–303.

Perron, O. (1940). Über lückenlose Ausfüllung des n-dimensionalen Raumes durch kongruente Würfel I & II, *Math. Z.* **46**, pp. 1–26, 161–180.

Peterson, W. W. (1961). *Error-Correcting Codes*, M. I. T. Press.

Pittel, B. (1984). On growing binary trees, *J. Math. Anal. Appl.* **103**, pp. 461–480.

Plesken, W. and Pohst, M. (1977). On maximal finite irreducible subgroups of $GL_n(\mathbb{Z})$. I. The five and seven dimensional cases, *Math. Comp.* **31**, pp. 536–551.

Plesken, W. (1985). Finite unimodular groups of prime degree and circulants, *J. Algebra* **97**, pp. 286–312.

Poyarkov, A. (2003). Master thesis, Moscow State University.

Poyarkov, A. (2005). Random packings by cubes, *Fund. Prikladnaya Matematika* **11**, pp. 187–196 (in Russian) (in English: Poyarkov, A. (2007). Random packing by cubes, *J. Math. Sci.* **146**, pp. 5577–5583).

Pyke, R. (1980). The asymptotic behavior of spacings under Kakutani's model for interval subdivision, *Ann. Probab.* **8**, pp. 157–163.

Pyke, R. and Van Zwet, W. R. (2004). Weak convergence results for the Kakutani interval splitting procedure, *Ann. Probab.* **32**, pp. 380–423.

Read, R. C. (1978). Every one a winner or how to avoid isomorphism search when cataloguing combinatorial configurations, Algorithmic aspects of combinatorics (Conf., Vancouver Island, B.C., 1976), *Ann. Disc. Math.* **2**, pp. 107–120.

Rényi, A. (1958). On a one-dimensional problem concerning random space-filling, *Publ. Math. Inst. Hungar. Acad. Sci.* **3**, pp. 109–127.

Robson, J. M. (1979). The height of binary search trees, *The Australian Computer J.* **11**, pp. 151–153.

Rogers, C. A. (1958). The packing of equal spheres, *Proc. London Math. Soc.* **3**, pp. 609–620.

Rogers, C. A. (1964). *Packing and Covering*, Cambridge University Press, Cambridge.

Schreiber, T., Penrose, M. D. and Yukich, J. E. (2007). Gaussian limits for multidimensional random sequential packing at saturation, *Commun. Math. Phys.* **272**, pp. 167–183.

Schrijver, A. (2005). New code upper bounds from the Terwilliger algebra and semidefinite programming, *IEEE Trans. Inform. Theory* **51**, pp. 2859–2866.

Schürmann, S. (2009). *Computational geometry of positive definite quadratic forms, Polyhedral reduction theories algorithms and applications*, AMS university lecture notes.

Shannon, C. E. (1948). A mathematical theory of communication, *Bell Syst. Tech. J.* **27**, pp. 379–423, 623–656.

Shepp, L. A. (1972). Covering the circle with random arcs, *Israel J. Math.* **11**, pp. 328–345.

Shepp, L. A. (1972). Covering the line with random intervals, *Z. Wahrscheinlichkeitstheorie und Verw. Gebiete* **23**, pp. 163–170.

Sibuya, M. and Itoh, Y. (1987). Random sequential bisection and its associated binary tree, *Ann. Inst. Stat. Math. A* **39**, pp. 69–84.

Sloane, N. J. A. (1984). The packing of spheres, *Scientific American* **1**, pp. 116–125.

Slud, E. (1978). Entropy and maximal spacings for random partitions, *Z. Wahrscheinlichkeitstheorie und Verw. Gebiete* **41**, pp. 341–352.

Solomon, H. (1967). Random packing density, *Proc. Fifth Berkeley Symp. Math. Stat. Prob.* **3**, pp. 119–134, Univ. of California Press.

Song, C., Wang P. and Makse H. A. (2008). A phase diagram for jammed matter, *Nature* **453**, pp. 629–632.

Stein, S. and Szabó, S. (1994). *Algebra and tiling (homomorphisms in the service of geometry)*, Carus Mathematical Monographs, **25**, Mathematical Association of America.

Szabó, S. (1986). A reduction of Keller's conjecture, *Period. Math. Hungar.* **17**, pp. 265–277.

Tanemura, M. (1979). On random complete packing by discs, *Ann. Inst. Statist. Math.* **31**, pp. 351–365.

Tanemura, M. and Hasegawa, M. (1980). Geometrical models of territory I. – Models for synchronous and asynchronous settlement of territories, *J. Theor. Biol.* **82**, pp. 477–496.

Thompson, T. M. (1983). *From error-correcting codes through sphere packing to simple groups*, The Mathematical Association of America.

Thue, A. (1910). Über die dichteste Zuzammenstellung von kongruenten Kreisen in der Ebene, *Norske Vid. Selsk. Skr.* **1**, pp. 1–9.

Torquato, S., Trusken, T. M. and Debenedetti, P. G. (2000). Is Random close packing of sphere well defined?, *Phys. Rev. Lett.* **84**, pp. 2064–2067.

Van Lieshout, M. N. M. (2006). Maximum likelihood estimation from random sequential adsorption, *Adv. Appl. Prob.* **38**, pp. 889–898.

Van Lint, J. H. (1975). A survey of perfect codes, *Rocky Mountain J. Math.* **5**, pp. 199–224.

Van Zwet, W. R. (1978). A proof of Kakutani's conjecture on random subdivision of longest intervals, *Ann. Prob.* **6**, pp. 133–137.

Widom, B. and Rowlinson, J. S. (1970). New model for the study of liquid-vapor phase transition, *J. Chem. Phys.* **52**, pp. 1670–1684.

Wiener, N. (1932). Tauberian theorems, *Ann. Math.* **33**, pp. 1–100.

Wilf, H. S. (1993). The asymptotic behavior of the stirling numbers of the first kind, *J. Comb. Theor. A* **64**, pp. 344–349.

Wilf, H. S. (2006). *Generating functionology*, A. K. Peters.

Wyner, A. (1964). Improved bounds for minimum distance and error probability in discrete channels, *Bell Telephone Laboratories Internal Report*, Murray Hill, N. J.

Zong, C. (1999). *Sphere Packings*, Springer-Verlag, New York.

Zong, C. (2005). What is known about unit cubes, *Bull. Amer. Math. Soc.* **42**, pp. 181–211.

Index